ESG Investment, German Industrie 4.0, and Blockchain

Kazuyuki Shimizu

ESG Investment, German Industrie 4.0, and Blockchain

Springer

Kazuyuki Shimizu
School of Business Administration
Meiji University
Chiyoda-ku, Tokyo, Japan

ISBN 978-981-95-6926-7 ISBN 978-981-95-6927-4 (eBook)
https://doi.org/10.1007/978-981-95-6927-4

This Springer imprint is published by the registered company Springer Nature Singapore Pte Ltd.
The registered company address is: 152 Beach Road, #21-01/04 Gateway East, Singapore 189721, Singapore

Preface

Background and Motivation

In recent years, the convergence of technological innovation, ESG investment, and digital governance has fundamentally redefined the foundations of capitalism. The emergence of Industrie 4.0, blockchain technology, and AI-driven analytics has transformed not only the nature of production and finance but also the ethical principles that sustain them. This book was written to examine these interconnections by integrating perspectives from economics, management, and philosophy—particularly focusing on how value creation can be reconstructed through the interplay of time, energy, and information.

Methodological Framework

As explained in the introduction, this book adopts an interdisciplinary approach that draws upon several prominent theories—including the Value Chain theory, Stakeholder theory, Corporate Governance theory, New Institutional theory, Actor–Network Theory (ANT), and Cybernetics—to interpret organizations as dynamic systems of learning and feedback.

In particular, it examines the industrial transformation that transcends the boundaries between the automotive and ICT industries in Japan, Germany, and the United States. The study integrates quantitative analyses (such as ESG investment data and EV adoption rates) with qualitative analyses that explore the ethical and social networks inherent in technological change.

Thematic Focus and Structure

The book is organized into four major parts:

1. Conceptual Foundations and ESG Investment,
2. Comparative Corporate Governance in Japan, Germany, and the United States—with particular reference to the case studies of Merck KGaA (Germany) and Merck & Co. (MSD, USA),
3. Integration of ESG and Industrie 4.0, and
4. Blockchain Technology and Digital Ethics.

Each part develops an independent yet interrelated argument, progressing from conceptual frameworks to empirical case studies and policy implications. The structure is designed to be accessible for both researchers and students, enabling its use as a university-level textbook as well as a reference for interdisciplinary research.

Intellectual Roots and Reflection

This study draws inspiration from thinkers such as Adam Smith, Max Weber, and Norbert Wiener, who each, in their era, confronted the tension between morality, productivity, and technology. For example, the discussion in the final chapter builds upon George Gilder's insight that "money is tokenized time," reinterpreting capital as stored and exchangeable time and examining the pluralization of value in the digital age through concrete cases such as Green Bonds utilizing IoT-based systems.

Acknowledgment and Outlook

The completion of this book owes much to the collaboration and insightful guidance of my colleagues and research partners in Japan, Germany, Ireland, and other European countries.

The discussions and reflections that took place during my Special Research Leave at Meiji University were closely connected to the broader globalization that has influenced academia since the 2000s. In Germany as well, a growing number of research presentations have come to be delivered in English, and major academic publications in Betriebswirtschaft (business economics) and management are increasingly written in English. These external developments have profoundly shaped both the conception and the evolution of this book. Moreover, my teaching experience during the intensive English-taught Block Week course at Osnabrück University in 2023 provided a significant inspiration for this book. In addition, the discussions and field studies conducted during my overseas research period served as a vital driving force that deepened both the theoretical and empirical aspects of the work.

Furthermore, the curiosity and critical thinking of international students at Osnabrück University and students at Meiji University offered invaluable intellectual stimulation, contributing greatly to the educational dimension of this book. Ultimately, this work aims to present a framework for understanding how technology and ethics coevolve, and to offer future generations the intellectual tools necessary to design a sustainable, decentralized, and human-centered digital economy.

Chiyoda-ku, Japan Kazuyuki Shimizu

Acknowledgments

This book could not have been completed without the generous support and encouragement of many mentors, colleagues, and family members. Above all, this research was made possible through the Special Research Program (Overseas Research) at Meiji University, approved and supported by the professors of the School of Business Administration. I am deeply grateful for the opportunity to pursue this study under this program.

I would also like to express my sincere appreciation to the members and organizers of the academic societies that have supported and inspired my research activities, including the Japan Association for Comparative Studies of Management, the Japan Academy of Business Administration, and the Japan Society for Information and Management. Their academic discussions and conference dialogues have provided invaluable opportunities for interdisciplinary collaboration and intellectual growth.

I wish to acknowledge with deep gratitude the members of my Ph.D. dissertation committee at the School of Commerce, Meiji University and my principal academic mentors, who have provided invaluable scholarly guidance and personal encouragement throughout the development of my research:

- Principal Examiner: Prof. Emeritus Nobutaka Kazama, Ph.D. in Commerce
- Co-Examiner: Prof. Kiyoshi Murata
- Co-Examiner: Prof. Hiroshi Yamashita, Ph.D. in Commerce and Ph.D. in Engineering

Although 18 years have passed since I obtained my doctoral degree in 2007, I remain profoundly indebted to my dissertation committee members—Professors Nobutaka Kazama, Kiyoshi Murata, and Yoji Yamashita—for their exceptional guidance and constant encouragement, which have continued to inspire and sustain my academic endeavors.

I would also like to take this opportunity to express my deepest respect and gratitude to the late Professor Emeritus Toshio Takahashi of the School of Business Administration, Meiji University, and Former President (14th Term) of the Japan Academy of Business Administration, for his invaluable guidance and mentorship both during my graduate studies and through our discussions at the IFSAM international conferences.

My sincere thanks are further extended to the following professors and colleagues, whose collaboration, insights, and friendship have greatly enriched my research through joint projects, academic discussions, and international cooperation:

- Prof. Michael J. Morley, Kemmy Business School, University of Limerick (Ireland).
- Prof. Dr. Johannes Hirata, Hochschule Osnabrück, Germany.
- Prof. Dr. Claudio J. Tessone, University of Zurich (UZH), Switzerland.
- Lecturer Dr. Hugo Pérez Moure, University of Vigo, Spain.
- Prof. Dr. Holger Bungsche, Kwansei Gakuin University.
- Prof. Dr. Robert Obermaier, University of Passau, Germany.
- Prof. Dr. Carsten Steinert, Hochschule Osnabrück, Germany.
- Prof. Dr. Julius Schöning, Hochschule Osnabrück, Germany.
- Prof. Dr. Takeshi Matsuda, Komazawa University.
- Prof. Takahiro Enjoji, Tokyo Fuji University.
- Prof. Dr. Masatsugu Fujie, Vice President and Professor, Faculty of Sociology, Urawa University, and Prof. Emeritus School of Business Administration, Meiji University.
- Prof. Hiroshi Koga, Kansai University.

I would like to express my sincere appreciation for the publication process. Finally, my deepest gratitude goes to my beloved family—my wife Michiko, my eldest son Tomoyuki, my daughter Miki, and my youngest son Masayuki—for their unwavering understanding and support during my extended research leave. Their patience, affection, and encouragement made it possible for me to devote myself wholeheartedly to research across different countries and disciplines.

To all who have contributed, directly or indirectly, to the realization of this book, I extend my heartfelt appreciation. This book represents not only the outcome of a personal intellectual journey but also the collective efforts and shared spirit of academic collaboration and lifelong learning.

Contents

List of Figures

List of Tables

Chapter 1
ESG Investment, German Industrie 4.0, and Blockchain: Conceptual Framework of This Book

Abstract The purpose of this chapter is to connect three domains—(1) ESG investment, (2) Industrie 4.0, and (3) blockchain technology—through the central concept of "data," and to illustrate how these elements are reshaping the structure of modern industrial transformation and corporate governance. The chapter also serves to organize the key concepts and theoretical frameworks necessary for understanding the four-part structure of this book.

First, the chapter redefines ESG investment by treating environmental, social, and governance factors not as mere numerical indicators but as "information" that gives meaning to decision-making. It also discusses the distinction between ESG and SRI, as well as the issue of greenwashing. Next, using the automotive industry as an example, it explains how IoT and Cyber-Physical Systems (CPS) in Industrie 4.0 reconfigure production and governance structures, thereby accelerating the datafication of corporate behavior. Furthermore, the chapter examines how the evolution of blockchain technology contributes to the reconstruction of trust, the development of decentralized governance, and greater transparency. Finally, it clarifies—through interdisciplinary perspectives from management studies, economics, and sociology—how these three domains complement one another as society unconsciously shifts (a phenomenon akin to the "reification" or "Fetischismus" of capitalism) toward a more sustainable socio-economic model.

Keywords ESG (Environmental, Social, and Governance) · Industrie 4.0 · Blockchain · Value chain · Stakeholder theory · Smiling curve concept

K. Shimizu, *ESG Investment, German Industrie 4.0, and Blockchain*, https://doi.org/10.1007/978-981-95-6927-4_1

1.1 Introduction

Digitalization collects, analyzes, and processes social interactions in real time, exerting a significant influence on the behavior of individuals and groups. In this dynamic process, the creative capacities of digital technologies give rise to various innovations, accompanied by both positive and negative societal impacts. Contemporary digital information and communication technologies (ICT)—including the Internet, artificial intelligence (AI), big data, and machine learning—are increasingly connecting individuals and social environments across national borders. This digital transformation is fundamentally reshaping social practices and affecting private life, economic activity, public engagement, and interactions with the natural environment.

However, in order for innovation to generate value for both people and the planet, appropriate governance mechanisms are indispensable. The development and deployment of digital technologies can lead to desirable or adverse outcomes, depending on the surrounding governance frameworks. Therefore, it is vital to investigate the nexus between digitalization and responsible innovation, and to assess how digital technologies can meaningfully contribute to sustainable development.

1.2 Conceptual Overview of This Book

This book provides an interdisciplinary examination of the interrelationship among three major concepts driving innovation in contemporary industry and finance—ESG investment (Environment, Social, Governance), Germany's Industrie 4.0, and blockchain technology. The content is structured into four main parts (see Fig. 1.1: "Conceptual Framework of This Study"):

- Conceptual Foundations and ESG Investment
- Corporate Governance and ESG Investment in Advanced Economies (Germany, the United States, Japan)
- Integrated Analysis of ESG Investment and Industrie 4.0
- The Development of Blockchain Technology

Although these themes appear to originate from different contexts, this book integrates them through the common thread of "data," analyzing both theoretical foundations and empirical cases. The introductory section outlines the theoretical basis of ESG investment, positioning it in relation to corporate social responsibility (CSR) and corporate governance theory. In parallel, it introduces analytical perspectives from value chain and "smiling curve" analyses in Industrie 4.0, thereby presenting a data-centered analytical framework. Furthermore, by applying the Stakeholder Matrix and the St. Gallen Management Model, the book constructs a stakeholder governance theory grounded in cybernetics.

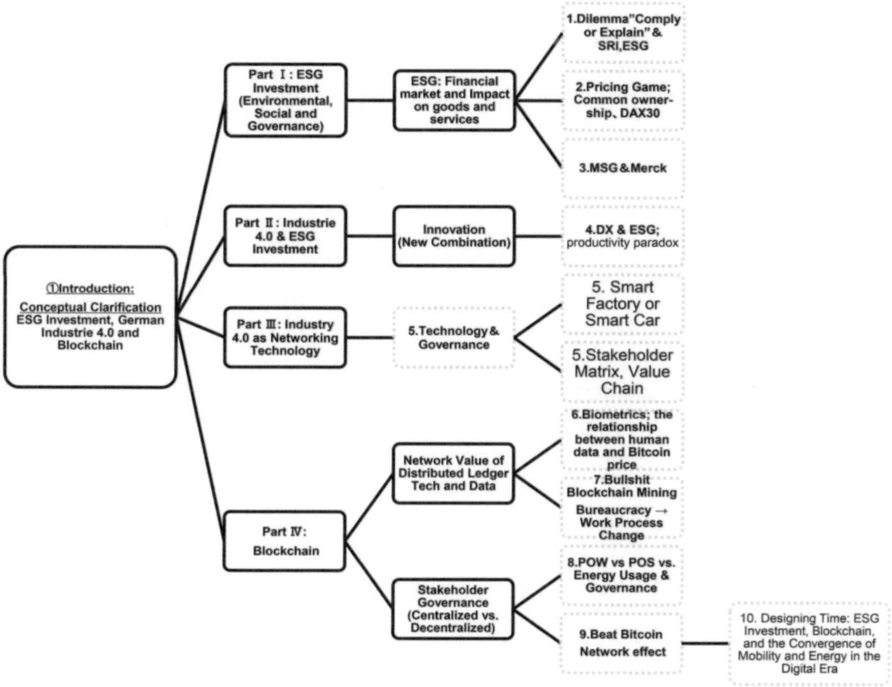

Fig. 1.1 Conceptual framework of this study. Created by the author

Part I: Redefining ESG Investment This section redefines ESG investment as follows:

- Environment (E) is reinterpreted in the context of energy transition,
- Social (S) is understood as a dynamic network among stakeholders, and
- Governance (G) is viewed as a vector that directs environmental and social networks.

Data essential for investment decisions should not be treated merely as numerical figures but as information containing meaning and context. Within the framework of pension fund management and blockchain integration, a new perspective of "the IoT-ization of human–asset management" is introduced.

1. "Comply or Explain" and The Methodological Dilemma Between SRI and ESG: Transitional Terminology; This chapter explains the development of Socially Responsible Investment (SRI) and Environmental, Social, and Governance (ESG) investing by comparing their concepts and methods. SRI is based on excluding certain industries for ethical reasons, while ESG focuses on transparency and information disclosure. The chapter reviews the historical shift from SRI to ESG and compares the performance of major sustainability indices such as FTSE4Good and the DJSI, including their behavior during financial crises.

It also introduces text-mining (LDA) results from annual reports of German automotive companies to illustrate how analytical methods are applied and what insights they provide. Finally, the chapter concludes that addressing greenwashing in ESG investing requires clearer investment concepts and the use of new technologies.

2. "Pricing Game" for Tacit Collusion and Passive Investment explains the structural factors behind tacit collusion in oligopolistic markets. As pricing becomes increasingly algorithm-driven, new forms of silent coordination between firms can emerge. Using game theory, the study analyzes how these interactions take place without explicit communication (Noria & Salima, 2004).

   The chapter also highlights how the rise of passive investors—particularly ETFs—reduces competition among firms because these investors often hold shares in multiple competing companies. Excess market liquidity and the growing complexity of asset management have accelerated this shift toward passive investment. As a counterbalance, the study suggests that family-owned businesses may help maintain healthier competition in the market.
3. Heterogeneity of Institutional Investors, Longevity, and Corporate Governance: The Case of Merck KGaA and Merck & Co.: This section compares Germany's Merck KGaA—whose governance is grounded in family capital and long-term stability, emphasizing "trust in people"—with the US Merck & Co., where trust is built through dispersed institutional ownership and the "immediacy of information." It explains how, in the Industrie 4.0 era, the basis of corporate trust is shifting from interpersonal relationships to mechanized reliability supported by data and algorithms (related papers: 1–4).

Part II: ESG Investment and Germany's Industrie 4.0 The Digital Transformation of Work and ESG as a New Framework for Social Security see Fig. 1.1, "4.DX & ESG; productivity paradox" explores how digitalization (DX) and the Internet of Things (IoT) are transforming labor structures, social security systems, and investment frameworks through a comparative analysis of Japan, Germany, and the United States. Although DX has improved productivity and reduced unemployment, it has also intensified inequality—the so-called "productivity paradox" (World Bank 2025). Within this transformation, ESG investment has evolved into a framework that reconciles social justice with economic sustainability. Moreover, blockchain-based auditing and data reliability are attracting attention as tools for preventing greenwashing. Each country pursues its own path toward balancing technology and social systems: Japan through welfare-oriented DX under Society 5.0, Germany through institutional reform under Industrie 4.0, and the United States through market-driven efficiency (related papers: 4).

Part III: The Intersection of ESG Investment and Industrie 4.0 "Technological Development in the Automotive Industry and Transformation in the Corporate Governance System," focusing on Germany's national strategy, Industrie 4.0, this part examines how IoT and Cyber-Physical Systems (CPS) influence corporate governance and the automotive industry. Case studies such as Volkswagen's modular production platform (MQB) and Toyota's TNGA illustrate how smart factories and smart cars are transforming production and distribution systems. AI-driven codification of tacit knowledge and autonomous M2M (machine-to-machine) communication foster a shift from "manufacturing" to "value creation." Data is positioned as the common foundation linking ESG and Industrie 4.0, and the discussion extends to the "Data Space" concept within Society 5.0, emphasizing real-time optimization and sustainability assessment (related papers: 5).

Part IV: The Expansion of Blockchain Technology Finally, this part analyzes the structural transformation of trust, value, and human relations driven by blockchain technology from four perspectives:

1. Biometrics Authorization and Blockchain—Explores how digitalization and the Internet of Body (IoB) affect human agency and decision-making, balancing privacy protection with ethical use of data.
2. Bullshit Blockchain Mining—Questions whether blockchain mining constitutes productive labor or a "bullshit job," while showing how decentralized networks may reconstruct social trust beyond bureaucratic limits.
3. PoW–PoS–DAO—Philosophically analyzes value creation through the contrast between Bitcoin (PoW) and Ethereum (PoS), using Hegel's dialectic and Popper's Three Worlds to explain the "countervailing power" of AI and DAO in harmonizing human and machine values.
4. Beating Bitcoin—Positions Bitcoin as "digital gold" and Ethereum as "digital oil," presenting a network-theoretical and power-law-based model of economic growth.
5. By linking blockchain, Industrie 4.0, and ESG investment, it identifies a paradigm shift from "Time is Money" to "Money is Time," symbolizing a new form of capitalized trust in the era of tokenized time (related papers: 6–10).

1.3 New Institutional Economics and ESG Investment

ESG investment is a concept recommended by the United Nations that incorporates Corporate Social Responsibility (CSR) into corporate governance. By considering a variety of stakeholders beyond shareholders, ESG investment aims to mitigate long-term corporate value risks and enable sustainable business management, expanding the traditional discussion of maximizing shareholder value.

Recent discussions on corporate governance have structured the concept of "stakeholders" into narrow and broad definitions (Freeman et al., 2001). Based on New Institutional Economics, the narrow definition primarily focuses on the relationship between "managers" and "shareholders." In contrast, the broad definition expands this framework by including other "stakeholders" on an equal footing with managers and shareholders (Shimizu, 2016 and 2017). Thus, discussions on corporate governance are closely related to CSR theory and can be structured based on their interrelationship. Corporate governance theory is explained from the perspective of three key theories in New Institutional Economics: 1. Transaction Cost Theory, 2. Agency Theory, and 3. The Theory of Property Rights. (Murray et al., 2021).

1.3.1 New Institutional Economics

1.3.1.1 Transaction Cost Theory

Transaction Cost Theory discusses the rationale behind choosing either "market" or "organization" based on transaction costs. From this perspective, ESG investment is believed to contribute to reducing transaction costs and improving risk management efficiency. Specifically, by enhancing corporate transparency, reducing long-term risks, and improving stakeholder relationships, ESG investment is expected to lower both direct and indirect transaction costs (Robert et al., 2014).

1.3.1.2 Agency Theory

Agency Theory highlights the potential conflict of interest between shareholders (principals) and managers (agents). Shareholders tend to prioritize dividends, whereas managers often focus on business growth, leading to occasional conflicts due to their differing interests. This conflict can be mitigated through ESG investment. Specifically, by aligning with stakeholders and implementing ESG strategies within the company, managers can reduce information asymmetry and improve transparency, thereby lowering the risk of a decline in stock prices and ultimately resolving the conflict of interest between the two parties.

1.3.1.3 Property Rights Theory

Property Rights Theory posits that clarifying ownership and attributing it to the rightful owners improves a company's economic efficiency and strengthens governance and stakeholder coordination. O. Hart emphasized the "Incomplete Contract Theory," pointing out that no contract is ever fully complete (Hart, 1995). In the case of incomplete contracts, company owners (mainly shareholders) exercise residual rights (decision-making rights over issues not covered in contracts) to impose ESG-related conditions on managers. Rational shareholders monitor managerial practices and encourage ESG-driven management, aiming to eliminate contractual incompleteness and improve overall corporate efficiency.

Thus, ESG investment expands the framework of corporate management by incorporating CSR elements that were previously overlooked due to the focus on maximizing shareholder value. It promotes more rational corporate management by reducing transaction costs, curbing agency costs between managers and shareholders, and improving management efficiency through the exercise of residual rights under Property Rights Theory.

1.3.2 *SRI (Negative Screening) and ESG (Positive Screening)*

Based on the institutional economics theories discussed so far,

Figure 1.2 illustrates the conceptual framework of this study (Concept 1) through two perspectives: SRI (Negative Screening) and ESG (Positive Screening). When comparing these two concepts, SRI (Socially Responsible Investment) reflects a restrictive approach that imposes constraints on economic activity through social and environmental considerations.

In contrast, ESG adjusts the interrelationships among environmental, social, and economic factors and introduces a more flexible, market-driven approach under the principle of "Comply or Explain." This results in a transformation of the conceptual framework.

1.3.2.1 Negative Screening in SRI (Socially Responsible Investment)

The concept of negative screening in Socially Responsible Investment (SRI) visually represents the relationship among the economy, society, and environment using concentric circles—where the environment encompasses society, and society encompasses the economy. This illustration conveys that economic activity is embedded within social structures, which in turn are subject to environmental constraints. In other words, economic activities are not independent but rely on both social frameworks and the sustainability of natural ecosystems (Shimizu, Technological Development in Automotive Industry and Transformation in Corporate Governance System, 2017).

1.3.2.2 Positive Screening in ESG Investment

In ESG (Environmental, Social, and Governance) investment, the three elements—environment, society, and economy—are regarded as independent concepts that are also interrelated. Within this framework, ESG aims to balance financial returns with

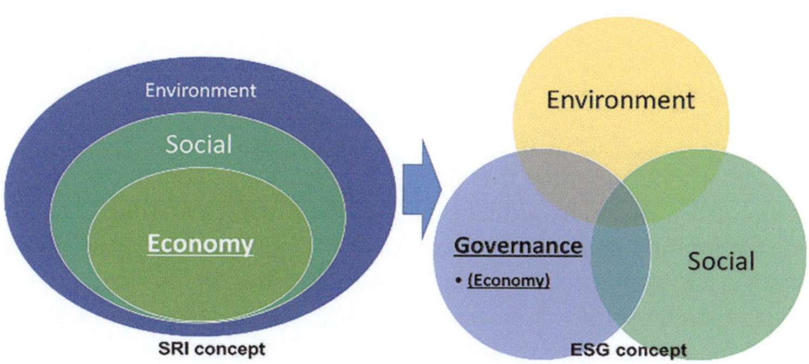

Fig. 1.2 Euler and Venn concept diagram. Created by the author

social and environmental impacts, thereby promoting the creation of long-term sustainable value.

Unlike traditional negative screening that excludes certain industries or behaviors, ESG investment emphasizes "positive screening," focusing on transparency and disclosure based on the "comply or explain" principle. This approach actively evaluates a company's environmental and social performance and encourages the practice of sustainable management.

The key difference between SRI and ESG lies in their decision-making approach and underlying values. SRI emphasizes ethical and social standards, adopting a negative screening strategy that excludes companies engaging in controversial practices. In contrast, ESG evaluates companies comprehensively through a balanced analysis of environmental, social, and governance factors, using a positive screening approach.

Put differently, SRI seeks to eliminate non-compliant companies, while ESG focuses on identifying and supporting firms that contribute to sustainability. ESG is designed to include a broader range of investable companies, aiming to generate long-term societal value on a larger scale.

1.4 Value Chain and Smile Curve Analysis

The connection point (②) in the conceptual framework of this study (Fig. 1.1), "Industrie 4.0 and Stakeholder Theory," is explained based on Porter's "Value Chain Model" and Shih's "Smile Curve Concept," shown in Fig. 1.4. In Germany, a unique program has been developed as part of the national economic policy on digitalization, promoting a network of embedded systems through the Internet, referred to as "Industrie 4.0." This program primarily targets the German automotive industry, aiming to establish competitive advantages across multiple countries through technological and organizational standardization. This concept of "connecting all industrial machines to the Internet" is positioned as part of the value chain in the national strategy, and aims to strengthen the competitiveness of the entire industry (Shih, 1996).

1.4.1 Value Chain

The value chain analyzes how a company connects and coordinates its business activities to create value in its products and services, while pursuing basic strategies such as differentiation and cost competitiveness—in other words, producing more efficiently or at a lower cost. From the perspective of strategic management, Michael Porter divides strategy into three hierarchical levels: corporate strategy, business strategy, and functional strategy, and positions the value chain as a framework that cuts across the latter two levels.

As part of functional strategy, the value chain is composed of primary activities and support activities. Primary activities include product development, production, marketing and sales, and customer service, all of which directly contribute to the creation of value. Support activities include the organizational infrastructure, information systems, procurement, and human resource management, which indirectly support and enhance the efficiency of the primary activities (Hill et al., 2020).

1.4.1.1 Primary Activities

The key components of the primary activities (see bottom of Fig. 1.3) are classified into four functions: research and development, production, marketing and sales, and customer service.

Research and Development

R&D (Research and Development) refers to activities related to product design and production processes. It encompasses the design of physical products—such as smartphones—as well as production methods in manufacturing firms, such as lean production. Importantly, service firms also engage in R&D. For instance, traditional banking systems are typically centralized. However, the advancement of ICT has enabled the development of decentralized payment systems through blockchain technology, which addresses the double-spending problem using competition-based data encryption. This has led to the creation of new financial products and markets utilizing various cryptocurrencies. Such systems achieve high levels of security and enable money transfers as easily as sending an email. (Nakamoto, 2008).

However, these innovations also come with challenges. Despite being potentially less secure, existing financial services such as online banking and smart debit cards

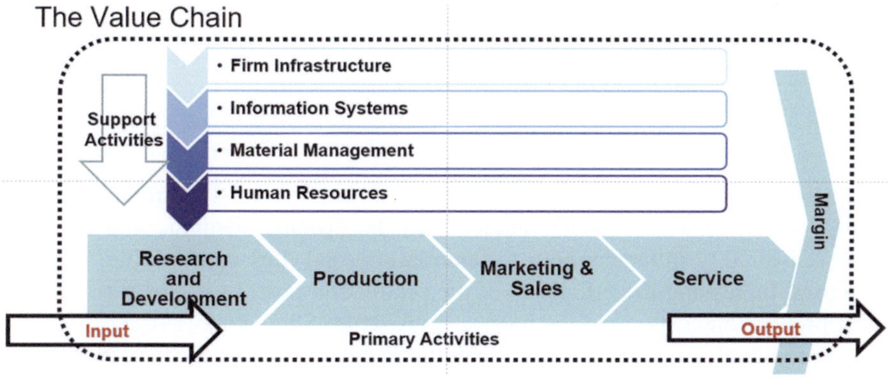

Fig. 1.3 Value chain model. Created by the author

continue to dominate in many aspects. This competitive tension arises due to several structural and societal factors, including:

- Regulatory uncertainty: Cryptocurrencies currently lack consistent alignment with traditional legal frameworks. As a result, institutional responses to risks like fraud or money laundering remain insufficient compared to regulated financial services.
- Heavy user responsibility: In decentralized systems, users must manage their own private keys. Losing access to one's key can result in irreversible loss of funds—an issue that traditional financial systems mitigate through support mechanisms such as password recovery.
- Volatility and the Oracle Problem: Cryptocurrencies exhibit high price volatility, which undermines their reliability as a medium of daily exchange. Moreover, blockchain systems face the Oracle Problem: they cannot independently verify real-world external data (e.g., weather, exchange rates). If incorrect data is input into smart contracts, this may trigger invalid or even harmful transactions.
- Infrastructure and cultural readiness: The current financial ecosystem has deeply rooted legal, commercial, and user experience systems. Replacing it with crypto-based alternatives requires not only technological development but also major institutional and cultural transformation.

R&D also plays a crucial role in reducing costs and enhancing product value. For example, the American semiconductor company NVIDIA has developed a powerful method of multimodal data processing through its proprietary chip technology. This allows the integration of not only text but also images and various forms of information, significantly accelerating the generation of big data and expanding the scope of machine-readable content—both of which are essential for the advancement of AI.

Moreover, R&D is not only about improving product functions or features, but also about considering the elegance of product design. A well-known example is Apple's iPhone, which has transformed electronic devices into fashion accessories. The elegant design of the iPhone reflects the aesthetic vision of its founder, Steve Jobs, who emphasized the concepts of "kallos" (beauty) and "graphein" (writing). This attention to design creates a strong connection between the product's value and the needs and desires of consumers.

Production

Production refers to the process of creating goods or services. The production function is closely tied to product quality, and the specific methods and activities employed directly influence both differentiation (and thus added value) and cost reduction. Increasing the efficiency of the production process makes it possible to lower production costs—and the reverse is also true (Shintaku 2016).

The foundation of modern production management can be traced back to Frederick W. Taylor, who introduced the principles of scientific management in the early twentieth century. Taylor's approach emphasized the analysis and standardization of work processes to improve labor productivity. His theories were later applied on a larger industrial scale by Henry Ford, who developed the Ford system of mass production. Ford's assembly line production method efficiently produced standardized goods in large volumes, greatly increasing output and reducing costs.

In contrast, the Toyota Production System (TPS) represents a post-Fordist model that emerged in response to changing market conditions, such as the era of low growth in the automotive industry. TPS enabled the production of a wide variety of products in small quantities. Its core concept lies in shifting the flow of production from a push-based system to a pull-based system, adapting flexibly to market demand. Known as "Just-In-Time," this approach minimizes waste by producing only what is needed, when it is needed, thereby enhancing efficiency and responsiveness (Shibagaki et al., 1989).

Marketing and Sales

A company's marketing and sales functions can generate value in a variety of ways. Marketing plays a key role in enhancing the perceived value of a product by influencing customer recognition, trust, and loyalty through brand positioning and advertising. In other words, marketing helps create a favorable image of the company's product in the minds of consumers.

In addition, securing customer loyalty can lead to new customer acquisition without the need for costly advertising. Recently, word-of-mouth marketing through social media has also become a highly effective promotional tool.

However, economist Thorstein Veblen took a critical stance toward this kind of value amplification through additional information, referring to it as the "dependency effect." For example, the French sparkling water brand Perrier sells at approximately 2.5 times the price of regular sparkling water, despite the fact that much of its value is attributed to brand image rather than the product itself.

From the perspective of the "smile curve," activities such as assembling computers are relatively simple and therefore generate little added value. As a result, many PC manufacturers enter the market, intensifying competition. In contrast, companies like NVIDIA add high value by designing advanced semiconductor chips. Although the actual manufacturing is outsourced to companies like TSMC, the design function itself is a high-value area with significant barriers to entry.

Moreover, the point at which value is added in a product's life cycle differs by industry. In the petroleum industry, for instance, there is little product differentiation in the gasoline sold at retail gas stations. However, substantial value is generated in upstream processes such as exploration, extraction, and refining, which require large-scale investment. These stages within the value chain account for a significant portion of the industry's value creation.

Customer Service

The role of a company's service function is to provide after-sales service and customer support. This function creates significant practical value by helping customers resolve issues and offering assistance after the product has been purchased.

For example, automobile manufacturers implement a wide range of after-sales services and support systems to maintain and enhance customer satisfaction after the sale of their vehicles. These services have evolved beyond simple repair responses to include preventive maintenance, emergency support, and remote services.

Typical services include scheduled maintenance and warranty programs. When purchasing a new vehicle, customers are commonly provided with complimentary inspections and extended warranty coverage, allowing for quick and efficient responses to mechanical failures or defective parts.

Emergency support services are also available 24/7. In the event of an accident or vehicle malfunction, customers can contact a call center or use a mobile app to request towing services or receive on-the-spot assistance.

Additionally, remote support through connected car services has become more prevalent. These systems continuously monitor vehicle conditions and automatically send alerts or schedule maintenance appointments when abnormalities are detected. This type of predictive service is gaining attention as a means of preventing breakdowns.

Furthermore, IoT-based services have been developed to support part replacements and recall management. Users receive notifications regarding worn-out components or product recalls, and necessary repairs are provided free of charge under a well-organized service system (Zhuming et al. 2025).

1.4.1.2 Support Activities

The support activities of the value chain provide inputs that allow the primary activities to take place. These activities are broken down into four functions: company infrastructure, information systems, materials management (or logistics), and human resources (see Fig. 1.3).

Company Infrastructure

A company's infrastructure refers to the overarching organizational foundation upon which all business activities are built. This includes several key elements:

- Organizational structure—how departments and roles are divided;
- Management systems—frameworks for performance evaluation and internal rules;
- Incentive mechanisms—systems for motivating employees through rewards;

- Corporate culture—shared values and approaches to work;
- Legal, accounting, and financial functions—specialized departments that support business operations.

Collectively, these components are sometimes referred to as the company's "organizational architecture" (Stewart, 1993). When properly designed and implemented, these systems enable a firm to deliver products and services effectively and efficiently.

In addition, the role of top management is crucial. Executives are responsible for designing and maintaining these infrastructural elements. Through strong leadership, they influence employee behavior and overall company performance. In this sense, top management itself can be viewed as an integral part of the corporate infrastructure.

A real-world example is Hiroshi Mikitani, CEO of Rakuten, who significantly transformed the company's infrastructure by implementing English as the official corporate language and introducing "Rakutenism," a set of core values and operating principles. These reforms reshaped the company's organizational culture, systems, and structure, enhancing its ability to compete globally. This illustrates how top management can actively shape and strengthen a firm's value creation platform.

Information Systems

An information system refers to a digital framework used to manage inventory, track sales performance, set prices, and respond to customer needs. In recent years, many companies have leveraged Internet-connected information systems to improve operational efficiency and enhance service quality. These systems enable organizations to more effectively manage their overall value-creation activities.

For example, UNIQLO has adopted RFID (Radio Frequency Identification) technology to streamline in-store operations such as inventory control and self-checkout. This has significantly reduced the time required for labor-intensive tasks like inspection and stocktaking, helping to prevent missed sales opportunities and potentially boosting revenue.

Moreover, this technology is not limited to individual stores. It allows real-time tracking and integration of information across the entire supply chain—including production, logistics, and sales—thereby contributing to overall operational efficiency and improving the accuracy of business planning throughout the company.

Materials Management (or Logistics)

The logistics function, also referred to as materials management, controls the flow of physical goods throughout the value chain—from procurement, through production, to distribution. The efficiency with which this process is executed can significantly reduce costs and, in turn, generate profits.

For example, the Spanish fashion brand ZARA offers over 10,000 diverse designs and produces them in small batches at its own sewing factories. Regardless of sales performance, ZARA intentionally limits the quantity of each item, creating a sense of scarcity and stimulating a consumer mindset of "buy it before it's gone." This approach enables the company to respond flexibly to changing fashion trends and to bring new items to store shelves within just a few weeks—an approach known as fast fashion.

Human Resources

The human resources (HR) function plays a crucial role in enabling companies to create greater value. Specifically, HR involves recruiting and assigning the right personnel, as well as maximizing their potential through training, motivation, and compensation systems. When human resources are managed effectively, employee productivity improves, costs can be reduced, and the quality of customer service is enhanced. As a result, the company can deliver more value to its customers.

In Germany, there is a unique system that significantly influences corporate HR practices, known as co-determination (Mitbestimmung). Under this system, employee representatives participate in company management. For large companies, German law requires that nearly half of the members on the supervisory board (Aufsichtsrat) be employee representatives.

This system makes it easier for companies to reflect employees' voices in their HR strategies, facilitating the development of fair compensation structures and more worker-friendly environments. As a result, employee motivation and loyalty increase, leading to higher productivity and improved customer service quality.

For instance, at the automobile manufacturer Volkswagen, employee representatives take part in the supervisory board. They collaborate with executives on decisions regarding employee development, working conditions, and factory safety measures. Such practices form a vital part of the company's HR infrastructure, supporting stable long-term management and high-quality production activities.

1.4.2 Smile Curve Analysis

The value curve represents the structure of value creation across an industry. The horizontal axis shows the industry flow, divided into three stages from left to right: upstream (component manufacturing), midstream (product assembly), and downstream (distribution and sales). The vertical axis indicates the amount of value added at each stage.

From a market competition perspective, the left side of the curve (upstream) is characterized by global competition, where success depends on technological capability, production efficiency, and economies of scale. In contrast, the right side (downstream) tends to involve regional or local competition, where brand strength, marketing channels, and logistics capabilities play a key role in competitive advantage.

In this context, the term "value" has two main dimensions:

1.4.2.1 Monetary Value

This refers to the economic value gained when a product or service is sold in the market—measurable in terms of profit, revenue, or price margins. For example, the profit margin gained by assembling components and selling them as finished products is a typical form of monetary value (Shih, 1996).

1.4.2.2 Value of Chainability

This refers to the potential for a product or service to connect and integrate with other offerings, creating extended or networked value. This dimension applies not only to the physical space (such as production and logistics) but also to the virtual space—including data, brand, and digital networks.

For instance, smartphones gain much of their value from how they are integrated with apps, cloud services, and social networks, forming a dynamic ecosystem. In this way, value is not self-contained, but amplified through connectivity.

This broader understanding of value goes beyond mere economic measurement. As French sociologist Jean Baudrillard noted, "A product is not simply the result of an industrial process, but a node where relationships, institutions, exchanges, and culture intersect."

In other words, a product is not only a physical item exchanged in the marketplace; it also carries symbolic, cultural, informational, and service-related dimensions. From this perspective, products are multi-layered entities that create not only economic value, but also social and cultural value in organized society.

This concept of the smile curve will be further developed in later chapters through case studies focusing on the automotive industries of Japan, the United States, and Germany. The discussion will explore how this framework applies to initiatives such as the "Smart Factory" and "Smart Car" in each country.

As previously described, the value chain consists of primary activities and support activities. In Fig. 1.4, the "smiling curve" is further illustrated by assigning labels to specific points along the horizontal axis, from left to right.

The areas labeled A, C, and E represent what is referred to as cyberspace. These segments involve the addition of high levels of intangible value, such as patents, technological innovations, and intellectual property, like software. In these domains, virtual contributions—rather than physical ones—drive value creation.

In contrast, the lower part of the diagram, labeled B, D, and F, represents physical space. This area corresponds to value added through the production and handling of physical goods. However, the potential for value creation in physical space is generally more limited compared to cyberspace.

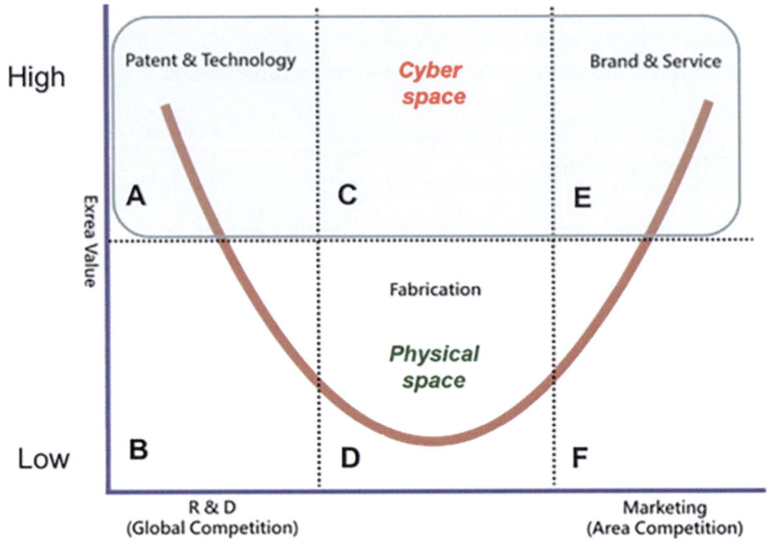

Fig. 1.4 Smile curve analysis. Created by the author

1.5 Stakeholder Matrix (Stakeholder Interests)

This paper adopts a cybernetic approach to analyze stakeholder governance. Cybernetics provides an effective framework for understanding the mechanisms of communication and control (Turnbull, 2012). Corporations are considered to be structured as a "nexus of contracts" (Jensen & Meckling, 1976), and cybernetics facilitates the sharing of information among stakeholders through ICT (Information and Communication Technology). This paper is structured into the following three sections:

First, it introduces an overview of the cybernetic approach.

Second, it examines how this approach can be applied to stakeholder governance.

Third, it analyzes 20 phenomena that emerge by applying the framework to five types of stakeholders and four environmental domains. Each section provides a step-by-step explanation from the perspective of social cybernetics, thereby deepening the understanding of stakeholder governance as an application of control theory. In addition, this approach incorporates both institutional aspects and practical elements of data science.

1.5.1 Luck Is Also a Part of Ability

Norbert Wiener's theory of cybernetics sparked the imagination for creating autonomous systems, such as humanoids. Cybernetics provides a framework for dealing with elements that are beyond direct control. For example, can "luck" be controlled?

Luck depends on individual perception. What is considered "good" varies from person to person (Norbert, 1950).

Luck, too, is part of ability. Wiener emphasized that the cybernetic perspective focuses on the relationship between animals and machines, and, in particular, on how a machine's internal mechanisms determine its performance and limitations. From this standpoint, the fact that the mechanical conditions of an insect constrain its intelligence is highly relevant. Based on this idea, the recognition of a specific situation—that is, the perception of "luck"—depends on the cognitive abilities of the perceiving individual.

This concept can also be applied to the theory of path dependence. Path dependence explains how past conditions and a sequence of decisions restrict current and future choices. David (1985) proposed three reasons for path dependence: technical interrelatedness, switching costs, and historical accidents (David, 1985). Here, we focus on the third element—historical accidents. Historical accidents are not only prerequisites but also components of ability. For example, examining the establishment of de facto technological standards reveals how various factors determine societal outcomes—such as the dominance of VHS over Betamax, or the competitive positioning of Windows against Apple in the operating system market.

Human visual recognition processes only about one ten-thousandth of the information captured by the eyes. Thus, the human perception of the external world is fundamentally based on connecting current sensory input with vast amounts of accumulated memory from the past to the present. In this context, activating the "default mode network, " which serves as a cognitive "space" that links sensory input with memory, becomes crucial for problem recognition.

Since cybernetics focuses on the relationship between animals and machines—particularly on how the internal structure of a machine determines its performance and limitations—it follows that, in humans, creating cognitive "space" to connect with vast memory reserves is essential for effective functioning. Wiener's theory highlights the commonalities between animals and machines and emphasizes that the internal structure of a machine determines its potential and limits. This perspective views AI not merely as a program but as a system shaped by its design philosophy and its relationship with the environment.

Recent advances in biology, such as human cloning and pluripotent stem cells, are enabling the reconstruction of biological "hardware." In response, AI can be seen as a "software entity" capable of being embedded into these new biological structures. Thus, we are entering an era where it is necessary to rethink the boundaries between artificial systems and organic systems. Even if AI and stem cell technologies have not yet fully matured, humans possess the adaptability to utilize and integrate available technologies as part of their environment.

This perspective offers important insights for stakeholder governance as well. It suggests that companies and societies must increasingly build strategies and institutions based on the premise of the co-evolutionary relationship among humans, AI, and machines.

1.5.2 Applying the Cybernetic Approach to Stakeholder Governance

To apply the cybernetic approach to the analysis of stakeholder governance, please refer to Fig. 1.5, which presents a conceptual framework. This Fig. 1.5 illustrates stakeholder governance, incorporating five key stakeholder groups. In the cybernetic approach, communication and interaction occur through distinct "modules" involving customers, suppliers, employees, financial stakeholders (such as shareholders, bondholders, and banks), and the community, all coordinated by the corporation (managed by professional executives). These relationships are visually represented in Fig. 1.5 (Ulrich, H. 1978).

1.5.2.1 Four Platforms

Each stakeholder is interconnected across four distinct platforms: the economy, technology, the natural environment, and society. These domains are based on the New St. Gallen Management Model (Rüegg-Stürm, 2005). However, this study does not adopt the hierarchical ordering of platforms found in the traditional St. Gallen Management Model from 1978. Instead, each stakeholder forms a framework centered around the corporation (managed by professional executives), and

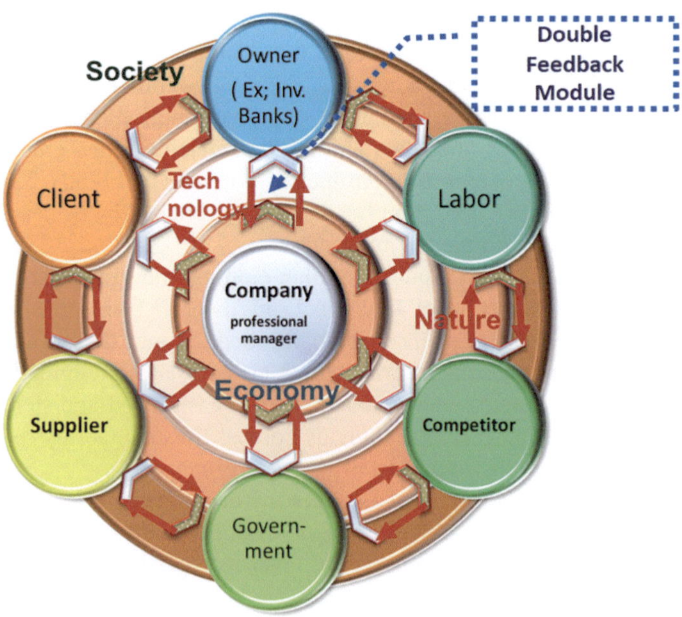

Fig. 1.5 Conceptual diagram of each of the stakeholders. Created by the author

the four platforms are organically linked within each module through a two-way exchange of information, commonly referred to as "double feedback."

Norbert Wiener pointed out that all intelligent behavior arises from feedback mechanisms, suggesting that human behavior could potentially be simulated by machines.

1.5.3 Stakeholder Mindsets

How, then, can these concepts be applied to stakeholder governance? Wiener's cybernetics can be applied to social organizations. For example, a governor regulates a market. This control system can be represented by the price system. In today's market economy, prices are determined by the forces of supply and demand. However, in this context, not only an economic perspective but also a managerial perspective is applied.

1.5.3.1 The Five Stakeholders and the Manager

The five stakeholders represent a rather narrow interpretation of the stakeholder concept. They are presented in sections A through E below.

Manager

The basic idea of creating value for stakeholders is extremely simple. A business can be understood as a series of relationships among groups who have a stake in the activities that make up the business. In other words, business itself is the ongoing process of how stakeholders and managers interact and create value together.

Customer

Customers exchange their own resources, such as money and time, for the company's products and services, and in return, they enjoy the benefits provided by those products and services.

Suppliers

Suppliers contribute to enhancing the company's value and are expected to build mutually positive relationships with the company. While some suppliers may focus primarily on price competition, maintaining fairness and transparency in relationships with suppliers is a crucial moral aspect.

Employees

Employees work for the company and typically depend on it for their livelihood. They expect a safe working environment, fair wages, benefits, and meaningful work. Many companies also offer employee stock-ownership plans, making employees financial stakeholders as well.

Financiers

Companies are surrounded by financial stakeholders, such as shareholders, bondholders, and banks. Owners or investors hold financial interests in the company through shares or bonds and expect a certain return on their investments. The ownership ratio and risk tolerance of financial stakeholders vary depending on the type of company and the characteristics of its investors.

Communities

Communities support businesses by granting permits for facility construction and providing essential resources for business operations. At the same time, businesses have a significant impact on the communities in which they operate. While creating value for other stakeholders, companies are also expected to act as good corporate citizens. Specifically, they must avoid causing unreasonable harm to the community, such as through pollution or the release of hazardous waste.

1.5.4 Four Environmental Domains in This Study

In the New St. Gallen Management Model, four key environmental domains are identified as having a significant impact on business activities:

1.5.4.1 Economy

The economic domain encompasses procurement, commercial activities, labor markets, and financial markets, forming the most crucial environment for corporate activities. Key aspects that businesses should pay attention to include the structure of both macroeconomies and microeconomies. Additionally, access to human resources, physical goods, capital, and information is vital, along with the efficiency of labor and financial markets and the availability of capital. It is also essential to understand the concentration levels of suppliers and customers. Furthermore, the state of public goods such as transportation and communication infrastructure must be carefully considered, as these greatly influence corporate operations.

1.5.4.2 Technology

Technological development serves as a vital source of competitiveness for companies. Key areas to monitor include biotechnology and genetic engineering, as well as process engineering that enhances manufacturing efficiency and material technologies that improve product performance. In addition, advancements in energy production technologies, transportation technologies, and robotics significantly affect corporate activities. Digital fields such as information and communication technology (ICT) and artificial intelligence (AI) have also become increasingly important in recent years. Moreover, the formation of regional technology clusters plays a major role in supporting corporate growth.

1.5.4.3 Natural Environment

For companies operating globally, natural environmental conditions are critical factors. Important points of focus include access to abundant resources such as air, water, land, mineral resources, and raw materials. Furthermore, the potential for the development of primary industries, such as agriculture, forestry, and fisheries, should also be considered. Climate conditions—such as temperature, humidity, and daily and seasonal variation patterns—are important when planning business activities. Moreover, companies must carefully assess biodiversity conditions and potential environmental pollution risks.

1.5.4.4 Society

Social conditions also have a major impact on corporate activities. Companies must be attentive to various social factors and their changing dynamics. Workforce productivity and education levels are key indicators directly tied to business growth. The degree of social acceptance of innovation and new ideas also strongly influences business development. Demographic structure, income and wealth distribution (such as measured by the Gini coefficient), and the occurrence of social issues or conflict risks are critical to ensuring business continuity. Additionally, attention must be given to the role of government, the process of public opinion formation, and the balance of political powers, as these factors affect social stability. Lastly, the development of public infrastructure and access to education are fundamental elements supporting business operations.

The feedback process within society is driven by the concept of "sympathy" proposed by Adam Smith. Smith suggested that individuals synchronize their thoughts and views through mutual sympathy when placed under similar conditions.

For example, companies like Google collect and analyze personal data to evaluate users' "levels of sympathy," optimizing the relevance and usefulness of search results accordingly.

Following Smith's theory, the presence of an "impartial spectator" is also assumed, representing a neutral perspective in evaluating actions.

Communication among these entities resembles machine-to-machine communication and is governed by a modular feedback system, as illustrated in Table 1.1.

1.5.5 The Integration of the Five Stakeholders and the Four Environmental Domains

In this section, we aim to apply the cybernetic approach to stakeholder governance while referring to the New St. Gallen Management Model. This approach addresses the following six key issues: (1) the four environmental domains, (2) the five stakeholders, (3) the interactions among stakeholders, (4) the structuring of power, (5) the processes, and (6) the modes of development.

The New St. Gallen Management Model was originally designed to reduce complexity in management systems. The specific functions of the 20 resulting factors are summarized in Table 1.1.

Particular attention should also be given to the role of the "COMPANY" itself, as it plays a critical role in stakeholder governance.

- The four environmental domains shown along the horizontal axis of Table 1.1 correspond to the four environmental spheres depicted in Table 1.1. These are:

  1. Economy,
  2. Technology,
  3. Nature, and.
  4. Society.

They represent the environmental perceptions that need to be controlled and managed.

Table 1.1 Stakeholder management: criteria and 20 elements

Criteria & 20 fators			Economy	Technology	Nature	Society
			1	2	3	4
Investor	A	COMPANY / Manager	1	6	11	16
Employee	B		2	7	12	17
Supplier	C		3	8	13	18
Customer	D		4	9	14	19
Community	E		5	10	15	20

Created by the author

- The five stakeholders, listed from A to E, are organized around the company (management) at the center of the conceptual diagram:

  A) Financial Stakeholders,
  B) Employees,
  C) Suppliers,
  D) Customers,
  E) Community.

  The vertical axis of Table 1.1 shows these five stakeholders.

Thus, combining the four domains (1–4) and the five stakeholders (A–E) results in 20 elements.

It is important to note, however, that these 20 elements represent individual factors and do not yet capture the interrelationships among them.

1.5.6 Theories of Diverse Stakeholders

"Luck" can also be considered a part of one's abilities. Norbert Wiener stated that "…the structure of a machine or an organism serves as an indicator of the performance expected from it."

If that is the case, then could "luck," understood as the anticipation of unforeseen events (fuzziness), be considered part of overall performance? It generates "variety" within a system.

Wiener expected that machines would deliver predictable performance and that specific capabilities could be ensured based on the material characteristics of their components.

This systems approach was designed to reduce complexity, supporting effective management practices. However, at the same time, it can disrupt the organic interactions among management systems themselves, thereby increasing the system's "variety."

1.5.6.1 Correlations Across Multiple Functions (Network Analysis and Deep Learning)

The relationships between the five stakeholders surrounding a company (customers, employees, suppliers, financial stakeholders, and the community) and the four environmental domains (economy, technology, nature, and society) are extremely complex. These relationships can be visualized using network analysis. In network analysis, "points" are referred to as nodes (actors or entities), and "lines" as edges (the flow of information or influence). By mapping nodes and edges, the structure of relationships becomes visible.

Furthermore, to predict and complement these complex connections, deep learning technologies are highly effective. Practical examples include Amazon's recommendation systems and knowledge supplementation models like ChatGPT, both of which combine network structures with AI.

By applying such methods to corporate governance analysis, it becomes possible to systematically understand multi-layered relationships—such as between companies, employment, employees, skills, and educational institutions—through the use of knowledge graphs.

In addition to strategically manage stakeholder governance, it is effective to organize the 20 relationships (5 stakeholders × 4 environmental domains) into a matrix, based on the St. Gallen Management Model.

Through this approach, companies are viewed as control targets within complex networks, promoting sustainable value creation.

Particularly important are three types of relationships:

- Human–Human (e.g., relationships among employees).
- Human–Machine (e.g., interactions with AI).
- **Machine–Machine** (e.g., autonomous communication between IoT devices).

Among these, the "Machine–Machine" relationship remains the most difficult to control, posing challenges such as malfunctions and the phenomenon of hallucination (false outputs generated by AI).

1.5.7 Institutional Perspective (Power, Legitimacy, and Urgency)

The procedure known as stakeholder analysis was specifically developed for this purpose (Mitchell et al., 1997). According to these proposals, the relative importance of stakeholders to an organization should be evaluated based on three dimensions: power, legitimacy, and urgency. Mitchell et al. (1997) identified seven stakeholder types, each possessing distinct levels of significance for the organization. The underlying idea is that the clearer the stakeholders are with regard to these three aspects—meaning the stronger, more legitimate, and more urgent their concerns—the more attention they should receive from the organization (Steinmann et al. 2013).

In this study, we conclude by introducing the concept of "variety." P. Hall and D. Soskice did not directly address our central theme of "A Cybernetic Approach to Stakeholder Governance." However, the interrelations between the four environmental domains and the five stakeholders can be understood as mechanisms for operating variety. Thus, we highlight the criteria and the 20 factors discussed earlier. We argue that ambiguity (unforeseen events) plays a crucial role in the performance of corporate governance systems and contributes to the creation of new, value-added activities.

1.6 Conclusion

This book offers an integrated analysis of three seemingly distinct areas: ESG investment, Germany's Industrie 4.0, and blockchain technology, focusing on their common underlying element—data.

Data is not simply numerical information; it also carries meaning, combining both quantitative aspects (such as investment amounts) and qualitative dimensions (such as applicability to sustainability). By redefining "data" as meaningful information, the book examines how innovations driven by ESG investment and Industrie 4.0 influence stakeholder perceptions and decision-making.

The main content is structured into four parts:

- Part I redefines ESG investment by emphasizing data not only as numbers but as meaningful, context-rich information essential for decision-making.
- Part II analyzes the transformative impact of Industrie 4.0 technologies, especially IoT and Cyber-Physical Systems (CPS), on the automotive industry through case studies of companies like Volkswagen and Toyota.
- Part III explores the intersection of ESG investment and Industrie 4.0, focusing on the construction of "data spaces" and real-time sustainability assessments.
- Part IV examines blockchain technology's role in enhancing decentralized governance, transparency, and security in zero-trust environments.

The theoretical foundation of the book combines New Institutional Economics (including Transaction Cost Theory, Agency Theory, and Property Rights Theory) with CSR theory, stakeholder theory, and cybernetic approaches. It proposes a framework for sustainable value creation and governance by emphasizing the dynamic management of data flows among multiple stakeholders.

Overall, this study aims to clarify how "data"—both as quantitative measures and as meaningful interpretations—acts as a central driver linking technological innovation and social transformation, thereby reshaping ESG investment and corporate governance models for the future.

1.7 Review Questions

1. Theories of New Institutional Economics and ESG Investment.

   Explain how the three theories of New Institutional Economics—Transaction Cost Theory, Agency Theory, and Property Rights Theory—contribute to understanding ESG investment.

   In your answer, discuss how each theory helps clarify the relationship between corporate governance and sustainable value creation.
2. Conceptual Difference Between SRI and ESG.

   Using the Euler diagram and Venn diagram presented in the text, explain the conceptual difference between Socially Responsible Investment (SRI) and Environmental, Social, and Governance (ESG) investment.

   How does each framework represent the relationship between the economy, society, and the environment?
3. Value Distribution in the Smile Curve.

   In the Smile Curve model, which stage—upstream, midstream, or downstream—creates the highest added value?

Provide specific industry examples to explain why value creation differs across these stages, and how technological or design innovation influences this distribution.

4. The New St. Gallen Management Model and Stakeholder Theory.

   Describe how the four environmental domains in the New St. Gallen Management Model—economy, technology, nature, and society—are connected to stakeholder theory.

   Explain how this model helps integrate environmental, technological, and social dimensions into modern corporate governance.

5. The five Stakeholders and four Environmental Domains Matrix.

   Explain the relationship between the five stakeholder groups and the four environmental domains introduced in the book.

   How do these 20 elements function as a matrix within the cybernetic framework of stakeholder governance, and what does this imply for corporate decision-making and control systems?

References

David, P. A. (1985). Clio and the economics of QWERTY. *The American Economic Review*. Retrieved from https://personal.utdallas.edu/~liebowit/knowledge_goods/david1985aer.htm

Freeman, E. R., Harrison, J., Wick, A., Parmar, B., & Colle, S. (2001). *A stakeholder approch to strategic management*. Blackwell.

Hart, O. (1995). *Firms, contracts, and financial structure*. Oxford University Press.

Hill, C. W., Gareth, J. R., & Melissa, S. A. (2020). *Strategic management: An integrated approach* (12th ed.). Cengage Learning.

Steinmann, H., Schreyoegg, G., & Koch, J. (2013). *Management: Grundlagen der Unternehmensführung Konzepte—Funktionen—Fallstudien Gebundene* (7 Ausg. ed.). Springer Gabler.

Jensen, M. C., & Meckling, W. H. (1976). Theory of the firm: Managerial behavior, agency costs and ownership structure. *Journal of Financial Economics*. https://doi.org/10.1016/0304-405X(76)90026-X

Mitchell, R. K., Agle, B. R., & Wood, D. J. (1997). Toward a theory of stakeholder identification and salience: Defining the principle of who and what really counts. *The Academy of Management Review, 22*(4). https://www.jstor.org/stable/259247?seq=1.

Murray, A., Kuban, S., Josefy, M., & Anderson, J. (2021). Contracting in the smart era: The implications of blockchain and decentralized autonomous organizations for contracting and corporate governance. Vol. 35, 4: Academy of Management Perspectives. doi:https://doi.org/10.5465/amp.2018.0066.

Nakamoto, S. (2008). *Bitcoin: A peer-to-peer electronic cash system*. Bitcoin.org. Retrieved from https://bitcoin.org/ja/bitcoin-paper

Norbert, Wiener. (1950). *The human use of human beings; Cybernetics and society*. Retrieved from http://21stcenturywiener.org/wp-content/uploads/2013/11/The-Human-Use-of-Human-Beings-by-N.-Wiener.pdf/2014.05.10

Noria, F., & Salima, H. (2004). *Managing computer networks security through self-organization: A complex system perspective*. ESOA: International Workshop on Engineering Self-Organising Applications. Retrieved 2018, from https://link.springer.com/chapter/10.1007/978-3-540-24701-2_9

Robert, E. G., Ioannis, I., & George, S. (2014). The impact of corporate sustainability on organizational processes and performance. *Management Science*. https://doi.org/10.1287/mnsc.2014.1984

Ronald, M. K., Bradley, A. R., & Donna, W. J. (1997). Toward a theory of stakeholder identification and salience: Defining the principle of who and what really counts. *The Academy of Management Review, 22*(4).

Rüegg-Stürm, J. (2005). *The new St. Gallen management model: Basic categories of an approach to integrated management*. Palgrave Macmillan.

Shibagaki, K., Trevor, M., & Abo, T. (1989). *Japanese and European management: Their international adaptability*. Univ of Tokyo Pr.

Shih, S. (1996). *Me too is not my style: Challenge difficulties, break through bottlenecks, create values*. Acer Foundation. Retrieved from https://laofutze.wordpress.com/wp-content/uploads/2010/07/me-too-is-not-my-style.pdf

Shimizu, K. (2016). Socio-cybernetic approach into the triumvirate: Stakeholder governance between management, shareholders and employees. An Enterprise Odyssey. *International conference proceedings*, 273-280. Retrieved from https://search.proquest.com/docview/1815362010?pq-origsite=gscholar&fromopenview=true

Shimizu, K. (2017). Technological development in automotive industry and transformation in corporate governance system. *Journal of Governance & Regulation, 6*(3). https://doi.org/10.22495/jgr_v6_i3_p5

Shintaku, J. (2016). Development of offshore production and its impact on domestic manufacturing. *Global Business Research Center: Akamon Management Review*. https://doi.org/10.14955/amr.150501

Stewart, R. (1993). *The reality of Organizations*. The Macmillan Press.

Turnbull, S. (2012). *A sustainable future for corporate governance theory and practice*. Principal: International Institute for Self-governance. Retrieved from https://papers.ssrn.com/sol3/papers.cfm?abstract_id=1987305

Ulrich, H. (1978). Unternehmenspolitik. Bern-Stuttgart.

World Bank. (2025, 4). *Manufacturing, value added (% of GDP)—North America, Japan, Germany, China.* Retrieved from https://data.worldbank.org/indicator/NV.IND.MANF.ZS?end=2023&locations=XU-JP-DE-CN&start=1991&view=chart

Zhuming, B., Chris, Z. W., Jianning, C., & Chaomin, L. (2025). *Internet of things (IoT) for sustainable mechatronic systems and cyber-physical systems (CPS).* ICINT2025.

Part I
ESG Investment (Environmental, Social and Governance)

Chapter 2
"Comply or Explain" and the Methodological Dilemma Between SRI and ESG: Transitional Terminology

Abstract This chapter examines the development of Socially Responsible Investment (SRI) and its relationship to ESG investing, focusing on how ethical screening and market-based accountability have shaped modern sustainable finance. SRI traditionally uses negative screening to exclude industries such as alcohol, tobacco, gambling, weapons, nuclear energy, and GMOs based on ethical and environmental concerns. ESG, in contrast, emphasizes transparency, disclosure, and the "Comply or Explain" principle introduced by the UK Cadbury Report, allowing firms to justify deviations from governance standards.

The chapter traces the historical foundations of SRI from Sharia-compliant finance and Methodist ethical investing to the creation of major sustainability indices such as the DJSI and FTSE4Good. It compares the performance of SRI and ESG using DJSI, FTSE4Good, and the MSCI World Index, applying Z-score standardization to evaluate differences between negative and positive screening.

In addition, text-mining (LDA) analysis of German automakers' annual reports is presented to show how EV adoption and national energy policies influence corporate disclosure and governance priorities. A review of global sustainable investment trends highlights that, while Europe and Japan have expanded their sustainable assets, the recent decline in US figures reflects stricter ESG definitions and efforts to eliminate greenwashing rather than a retreat from sustainability.

Finally, the chapter concludes that addressing greenwashing in ESG investing requires clearer investment concepts and the use of new technologies.

Keywords ESG (Environmental, Social, and Governance) · Exclusionary criteria · Mainstream investment strategies · Passive investment · SRI (Socially Responsible Investment)

K. Shimizu, *ESG Investment, German Industrie 4.0, and Blockchain*,
https://doi.org/10.1007/978-981-95-6927-4_2

2.1 Introduction

This chapter explores the methodological and conceptual differences between ESG (Environmental, Social, and Governance) investment and SRI (Socially Responsible Investment). While these two frameworks emerged from distinct historical trajectories, a major turning point occurred with the publication of the Cadbury Report in the UK in 1992. This report introduced the principle of "Comply or Explain," requiring companies not only to follow governance codes but also to provide clear explanations when they deviate. This principle reshaped the framework for evaluating firms and investment decisions, laying the groundwork for the rise of ESG investing.

Against this backdrop, the contrast between SRI—focusing on ethical values, such as Islamic finance—and ESG—emphasizing social accountability and market efficiency—has become a key concern among investors. Notably, by 2022, the cumulative volume of ethical investment in Europe exceeded that of the United States, where shareholder activism is more common. This trend reflects the stricter US criteria for sustainable investment, which reduced nominal totals but arguably strengthened ethical standards.

SRI is characterized by negative screening, which excludes ethically problematic industries (e.g., arms, tobacco, gambling). ESG, in contrast, adopts a positive screening approach: even firms in such industries may be included if they offer sufficient social accountability. This represents a shift from rule-based exclusion to market-based evaluation.

At the core of ESG lies the "Comply or Explain" principle, whereby investment decisions rely on transparent disclosures and accountable governance. This framework aims to reconcile ethical concerns with market logic. However, ESG investment still faces challenges—such as "greenwashing," where companies falsely claim environmental friendliness, as seen in the 2015 Volkswagen emissions scandal. Such externalities are difficult to capture through market pricing alone. Therefore, this chapter examines the methodological dilemma between ESG and SRI and offers guidance on addressing economic externalities like greenwashing.

In this context, the use of blockchain technology—which enables real-time, transparent monitoring of environmental and social performance—is gaining attention as a way to combine the strengths of both ESG and SRI. This technology facilitates a hybrid investment model that allows for organizational ambidexterity—the capacity to balance ethical constraints and market adaptability across conceptual boundaries.

To address these dilemmas, this chapter proceeds along three analytical dimensions:

1. **Conceptual Clarification: Negative vs. Positive Screening**

   Understanding SRI and ESG benefits from diagrammatic comparison. SRI adopts a nested model, where economics is embedded in society, and society in the environment—emphasizing the finite nature of resources. ESG, by contrast, presents these dimensions as intersecting, highlighting mutual interaction but not explicitly framing resource constraints.

While SRI relies on ethical exclusion, ESG presumes accountability and inclusion. With the rise of blockchain technology, a new hybrid model is emerging—capable of monitoring entire supply chains in real time and integrating both ethical and market-based assessments.

2. **Screening Strategies and Evaluation Methods**

According to the 2022 report by the Global Sustainable Investment Alliance (GSIA), the total assets under sustainable investment strategies—whether negative or positive screening—had nearly doubled from 2016, reaching USD 30.3 trillion.

Many ESG index providers exclude sectors such as alcohol, gambling, tobacco, weapons, firearms, nuclear energy, and GMOs. Between 2020 and 2022, the United States applied stricter standards, excluding ESG assets that lacked sufficient disclosure. These exclusions reflect diverse ethical, religious, and cultural norms. SRI, in particular, has roots in values-based systems such as Islamic finance (Sharia).

By 2025, ESG interpretation has evolved through advances in ICT:

- Environment (E) is now defined by energy transition and carbon neutrality.
- Social (S) is viewed as stakeholder networks (employees, customers, communities).
- Governance (G) functions as the directional vector connecting E and S.

Furthermore, AI has begun to redefine "data" not merely as numbers, but as context-rich information, prompting a reassessment of what counts in investment decisions.

3. **Performance Analysis: SRI/ESG Metrics and Text Mining.**

Quantitative assessment of SRI and ESG performance has advanced through AI tools (e.g., ChatGPT) and statistical programming (e.g., Python). This textbook compares three major indices (using 2016 data):

- MSCI World Index (traditional global stock benchmark),
- DJSI (Dow Jones Best-in-Class Indices formerly Dow Jones Sustainability Index, the first ethical investment index),
- FTSE4Good (aligned with the UK's "Comply or Explain" principle).

Findings show a high correlation (0.9743) between MSCI and FTSE4Good during periods of market turmoil, suggesting that differences between negative and positive screening were not yet distinctly reflected in stock selection. While some prior studies suggest that ethically sound firms offer long-term risk mitigation and stability, further research is needed on how screening strategies shape financial outcomes (Mans-Kemp, N. 2014).

This chapter also introduces text-mining methods, particularly Latent Dirichlet Allocation (LDA), applied to annual reports from Germany's major automakers. LDA identifies latent topics within qualitative/quantitative documents and maps stakeholder relationships within sustainable investment contexts.

This allows analysts to differentiate between linear, financially oriented "sustaining innovation" and exponential, "disruptive innovation" captured in environmental discourse—providing insight into firms' organizational responsiveness. Technologies like blockchain can enhance this by enabling rapid feedback, evolving investment behavior toward hybrid models that transcend traditional conceptual boundaries.

2.2 What Is SRI?

SRI refers to investment strategies that incorporate ethical, social, and environmental criteria. Its goal is not only financial return, but also corporate responsibility and behavioral change. Unlike traditional shareholder activism, SRI engages companies through shareholder rights to address social and environmental issues (Crane & Matten, 2010).

> SRI prioritizes sustainability by including governance and non-financial risks in its evaluation. In this sense, SRI aligns with the vision of a green economy, where economic activity is embedded in social networks and bounded by environmental constraints (Cowton, 2018).

According to Deutsche Bank, global financial assets total approximately USD 294 trillion. Thomas Piketty (2014) argues that if the return on capital (r) exceeds economic growth (g), wealth concentration leads to social and economic instability. Against such dynamics, sustainable investment had reached USD 30.3 trillion globally by 2022. Europe grew from USD 12 trillion (2020) to USD 14 trillion (2022); Japan from USD 2.9 trillion to USD 4.3 trillion. Conversely, the US numbers fell from USD 17.1 trillion (2020) to USD 8.4 trillion (2022).

However, this decline does not imply a retreat from sustainable investing. Rather, it reflects a maturing of ESG definitions and reporting standards, particularly efforts to eliminate greenwashing.

In 2022, the US SIF excluded assets that lacked clear ESG criteria—such as human rights or biodiversity considerations—even if they claimed ESG alignment. Additionally, overlapping strategies were disallowed, reducing reported AUM. These changes reflect a broader international trend toward greater transparency and stricter disclosures (Marcos & Vanessa, 2024).

Thus, the US decline signals not a contraction but a refinement of ESG metrics. Going forward, international comparisons must account for definitional and reporting differences across regions, see Fig. 2.1.

2.2.1 *What Is an Index?*

Stock prices are fundamentally determined by the relationship between supply and demand. If there are more investors (buyers) wanting to buy a particular stock, the price rises as demand increases. Conversely, if more investors want to sell, prices fall as sellers attempt to sell quickly, even at lower prices. In this way, stock prices are dynamically formed based on the power balance between buyers and sellers.

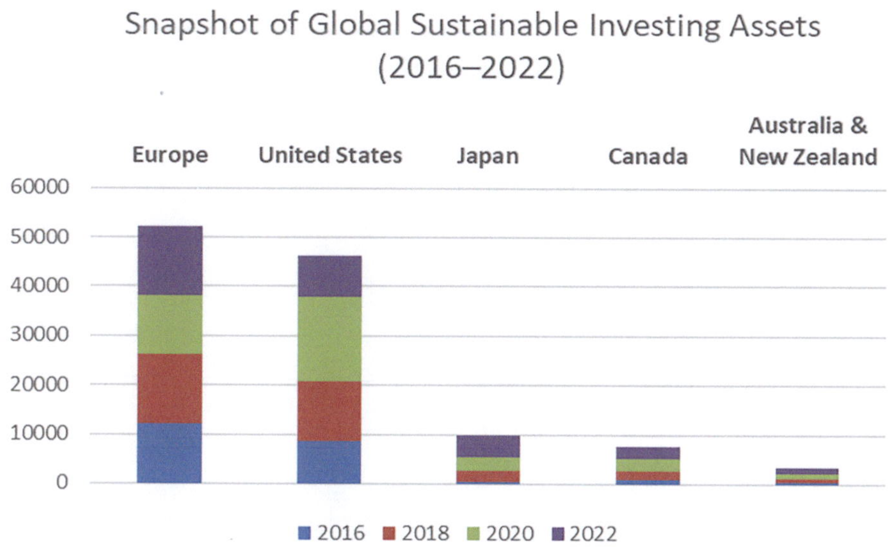

Fig. 2.1 Sustainable investment asset trends. Source: The Global Sustainable Investment Allience (GSIA) (2016, 2022), Created by the author

An index (stock market index) captures the movements of individual stocks from a macroeconomic perspective. For example, the US S&P 500 is based on the stock prices of 500 leading American companies, while Japan's Nikkei 225 consists of 225 representative stocks. These indices are widely used as indicators for understanding overall economic trends and for investment decisions.

Indices also play a crucial role in risk management. The investment proverb "Don't put all your eggs in one basket" highlights the importance of portfolio diversification across different assets. Index-based investment naturally achieves diversification by investing in a basket of companies.

There are two major investment strategies in the stock market: active investment and index investment. Active investment seeks to outperform the market index by selecting specific stocks and determining the timing of transactions, requiring a high level of expertise and incurring higher fees. On the other hand, index investment involves investing in funds or ETFs designed to track specific indices, requiring less effort in stock selection and transaction timing, and offering low costs and high diversification. As a result, it has gained popularity as a stable, long-term investment strategy.

2.2.2 *SRI (Negative Screening) and ESG (Positive Screening)*

2.2.2.1 Model of Sustainable Investment

As shown in Table 2.1 "Conceptual History of SRI and ESG," SRI is rooted in the ethical principles of Islamic Sharia law. In the United States, the concept gained traction after the 1929 stock market crash and the Great Depression of the 1930s,

Table 2.1 Historical development of SRI and ESG concepts

<table>
<tr><th>Methodological Trend</th><th>Year</th><th>Movement</th><th>What?</th></tr>
<tr><td rowspan="3">Pure exclude strategy</td><td>1700</td><td>Origin: SRI Fund</td><td>Islamic Sharia-compliant investing is inherently based on moral values. Similarly, the Methodist movement, inspired by John Wesley, practiced exclusionary investing by avoiding unethical industries such as alcohol, tobacco, gambling, and pornography.</td></tr>
<tr><td>1928</td><td>Pioneer: SRI Fund</td><td>A religious group in Boston pioneered the first exclusionary Socially Responsible Investment (SRI) fund.</td></tr>
<tr><td>1960s</td><td colspan="2">"Silent Spring" by Rachel Carson, published in 1962, along with the Civil Rights Act of 1964 and the Voting Rights Act of 1965, marked a turning point in American social and environmental consciousness.</td></tr>
<tr><td rowspan="3">Inclusion strategy (stakeholder Approach)</td><td>1970s</td><td colspan="2">Corporate managers were required to reconsider the social responsibilities of business enterprises in the 1970s. Lobbying activities were particularly effective in advancing human rights—For example, through affirmative action and divestment campaigns related to apartheid in South Africa.</td></tr>
<tr><td>1980s</td><td colspan="2">A nuclear environmental disaster occurred in Chernobyl in 1986, in what was then part of Ukraine under the Union of Soviet Socialist Republics (USSR). As a result, the screening process for socially responsible investing needed to expand beyond basic exclusion criteria—Such as weapons, alcohol, tobacco, and gambling—To also include environmental and energy-related issues, such as nuclear power, environmental pollution, and labor conditions.</td></tr>
<tr><td>1984</td><td>Stakeholder approach</td><td>Freeman, E., A stakeholder approach to strategic management</td></tr>
<tr><td colspan="4">↑SRI "comply or explain" ↓ESG</td></tr>
</table>

Comply or explain (ESG approach)	1992	The Cadbury report	It sets out recommendations on the structure of company boards and accounting systems to mitigate risks and failures in corporate governance.
	1999	DJSI world	The Dow Jones adopts a best-in-class approach that enables the selection of sustainability leaders across all sectors, without discriminating against organizations from any particular industry.
	1998	German KonTraG	Control and transparency in business (KonTraG)
	2001	FTSE4GOOD	FTSE develops responsible indices designed to track the performance of companies exhibiting robust environmental, social, and governance (ESG) practices, utilizing an exclusionary screening methodology.
	2002	USA federal law	Sarbanes-Oxley act
	2006	The Principles for Responsible Investment (PRI)	The Principles for Responsible Investment (PRI), developed by institutional investors, aim to integrate environmental, social, and governance (ESG) considerations into investment decision-making. In doing so, institutional investors must also uphold their fiduciary responsibilities toward their beneficiaries.
	2008	Financial crisis	Financial crisis in USA (Dow from 12,500p to 7000, and back to 22,000p in 2017)
	2010	The Dodd–Frank Wall street reform	The Dodd–frank act was enacted in the U.S. after the 2008 financial crisis to strengthen financial regulation, improve transparency, and ensure accountability.
	2010 (1990)	MSCI ESG (Domini)	MSCI established a leading position in ESG evaluation by acquiring RiskMetrics group, which owned KLD research—The firm behind the Domini 400 social index that applied both positive and negative screening.

Source: Researcher's own construction based on several studies

which led to the standardization of financial reporting. Both SRI and ESG emphasize the importance of corporate disclosure that considers stakeholder interests.

The logical structure of SRI is depicted in Fig. 2.2's model, which consists of three concentric segments: the innermost economic segment, surrounded by the social segment, and finally, the environmental segment that has gained attention since the 1960s–1970s. This structure reflects the principle of inclusion in SRI.

The diagram conveys a clear stance on environmental resource limits—economics exists within society, and society exists within nature. This suggests that both the economy and society depend on environmental constraints. For instance, pollution from nuclear power plants is an externality of economic activity that must be addressed by society and environmental segments.

This understanding—where economic activity takes place within social networks, and society exists within the natural environment—forms the foundation of SRI and highlights the importance of making investment decisions based on environmental limitations.

Conventional economics focuses on monetary value and economic growth. However, the growing seriousness of environmental and social issues has brought attention to the concept of the green economy, which positions the economy within society and society within the environment. This triple-layered structure emphasizes diversity, interconnectedness, and embeddedness, redefining economics as an activity rooted in social relations and emphasizing harmony and coexistence with nature and future generations (Parkin et al., 2003).

SRI and ESG investment are extensions of this philosophy. SRI generally uses a hierarchical structure (environment > society > economy) and employs negative screening to exclude companies involved in tobacco, weapons, fossil fuels, etc., based on ethical and resource considerations. ESG, on the other hand, sees the elements of Environment (E), Social (S), and Governance (G) as overlapping and

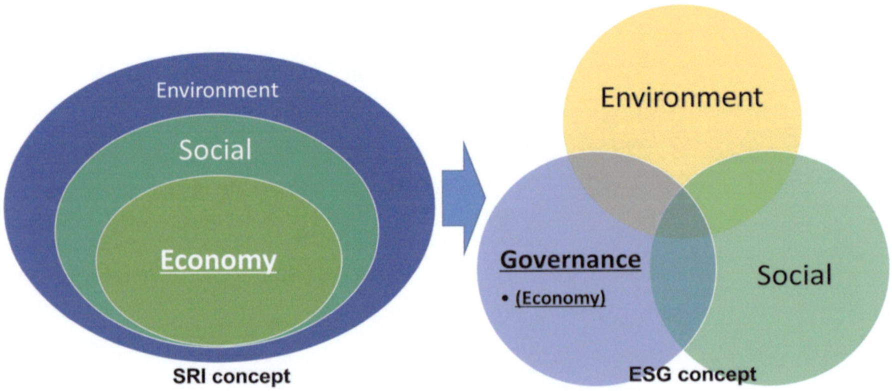

Scott Cato, M. (2009). Green Economics. London: Earthscan, p. 37.

Fig. 2.2 Concepts of SRI and ESG (Created by the author)

interconnected, focusing on transparency and accountability (Comply or Explain), and uses inclusive, positive evaluation criteria.

More recently, blockchain technology has enabled greater transparency and traceability in supply chains, leading to hybrid investment models that combine SRI and ESG features—offering both moral accountability and market efficiency. In this paradigm shift from growth-centered to coexistence-centered thinking, investment becomes an act of discovering new value within the sustainable relationship among economy, society, and environment (Scott, 2009; Knoll, 2002).

2.2.2.2 A Brief History of SRI

The concept of SRI compliant with Sharia law is influenced by eighteenth-century Methodist movements in the West, which emphasized moral values. These thinkers screened investments to avoid companies involved in alcohol, tobacco, gambling, and pornography. Empirical studies suggest SRI tends to underperform traditional active investment (Poelloe, 2010). Green mutual funds show lower returns but similar risks compared to conventional mutual funds, underperforming on a risk-adjusted basis (Chang et al., 2012). Barnett and Salomon (2002) found that financial returns initially decrease with more social screens but recover when the number of screens peaks—indicating a curvilinear and potentially complementary relationship (Barnett & Salomon, 2002). However, the empirical dilemma remains under-researched.

- c. 700 CE: Emergence of Sharia-compliant ethical finance,
- 1700s: Methodist movement promotes moral investing
- 1928: First negative screening fund launched by a Boston religious group (Blaine, 2020)
- 1992: The Cadbury Report introduced the concept of "Comply or Explain." This principle reflects a clear market-driven approach: if investors do not accept a company's explanation, they are free to sell their shares (Hirschman, A. O. 1970). The official title of the report is Report of the Committee on the Financial Aspects of Corporate Governance, published in 1992 by a UK committee chaired by Sir Adrian Cadbury (The Committee on the Financial Aspects of Corporate Governance, 1992). The report focused on improving corporate governance by recommending clearer allocation of responsibilities within the board, enhanced transparency, and especially the establishment of more independent supervisory structures through nonexecutive directors (see midsection of Table 2.1) (Yulia, 2017)
- 1999: Dow Jones launched the Dow Jones Sustainability Index (DJSI), which has since become a leading benchmark in the field of sustainable investment.
- 2001: With the expansion of SRI across Europe, the UK launched the FTSE4Good Index in 2001.
- Early 2005: Then-UN Secretary-General Kofi Annan called on some of the world's largest institutional investors to participate in the development of the Principles for Responsible Investment (PRI).

- 2010: MSCI acquired RiskMetrics Group, which owned KLD Research. KLD Research had developed the Domini 400 Social Index (The Domini Index 2026), launched in 1990, which became one of the first major benchmarks in socially responsible investing.

2.3 SRI and ESG Investment Approaches

2.3.1 SRI's Negative Screening vs. ESG

In this chapter, SRI is straightforwardly defined as a negative screening approach, while ESG represents a positive screening methodology. However, in practice, it is often difficult to clearly distinguish between these two investment policies. When examining sustainability-focused funds that incorporate both SRI and ESG perspectives, one finds that actively managed funds and equity funds dominate the top 50 ESG investment portfolios. This suggests that companies with ESG considerations tend to deliver attractive equity performance for investors (The Global Sustainable Investment Alliance (GSIA), 2016, 2022, 2025).

Firms with high ESG scores have consistently shown strong medium- to long-term performance. However, this does not imply that "all ESG-compliant firms generate high returns"—rather, it reflects selective investment in companies that integrate ESG effectively.

Within ESG investment strategies, negative screening—particularly within actively managed equity funds—excludes specific industries or companies based on ESG standards or social norms. This approach is deeply rooted in ethical, religious, and cultural values. SRI (Socially Responsible Investment), in particular, draws upon religious frameworks such as Islamic finance (Sharia). For instance, the Eurosif (European Sustainable Investment Forum) identifies the following industries as common exclusion targets under negative screening strategies:

- Alcohol
- Gambling
- Tobacco
- Weapons and civilian firearms
- Nuclear energy
- Genetically modified organisms (GMOs)

As of 2025, the concept of ESG (Environmental, Social, Governance) is being dynamically redefined through the rapid advancement of ICT (Information and Communication Technology). Each component of ESG has acquired new implications:

- Environment (E): Emphasis is placed on decarbonization and the shift toward renewable energy, with climate change measures now forming the core of corporate strategies.

- Social (S): Increasing focus is given to how companies build relationships with employees, customers, and local communities—in short, stakeholder networks.
- Governance (G): Governance functions as the "vector" connecting E and S, emphasizing transparency in decision-making and the effectiveness of corporate accountability.

This reconfiguration is significantly influencing investment decisions, especially those employing negative screening strategies. Negative screening, rooted in ethical and social values, excludes industries such as alcohol, tobacco, gambling, weapons, nuclear power, and GMOs from investment targets. The process emphasizes the proportion of revenue or profits derived from these controversial products or services. For example, if a company's revenue from tobacco-related products exceeds a certain threshold, it is excluded from investment portfolios.

To quantify the effect of exclusions, the valuation weights of these so-called "vice stocks" (industries considered unethical) are reduced relative to standard free-float-based valuations.

Investors highly committed to ESG engagement are increasingly proactive in influencing corporate strategy through proxy battles, thereby reinforcing a direct impact on managerial direction.

2.3.2 Controversial Industries Under Strong Exclusion

Industries such as alcohol, tobacco, gambling, military weapons, nuclear energy, and genetically modified organisms exert a powerful influence over industrial society (Matthews, C. 2016). Even when excluded from ESG-focused investments, these industries often retain significant lobbying power in political institutions and continue to satisfy key stakeholders.

Norges Bank, adhering to ethical guidelines and based on international standards such as those of the EU, the United States, and UN, closely monitors corporate lobbying and involvement in public policy. When concerns arise, they initiate a stepwise process of dialogue → observation → exclusion, aimed at reducing long-term investment risk while preserving moral credibility (Norges Bank, 2020).

2.3.2.1 Alcohol

According to the World Health Organization (WHO), harmful alcohol consumption is considered one of the major global health risks. It is a causal factor in more than 60 major diseases and injuries and is responsible for approximately 2.5 million deaths each year. The global alcohol beverage market has reached an estimated scale of USD 1.7–2.5 trillion (compared to USD 1.2 trillion in 2014) and is projected to grow at an annual rate of more than 5% over the next 5 years (Fortune Business Insights, 2025). See below Table 2.2 top three companies in the world.

Table 2.2 Top three global alcohol companies (2025)

Rank	Company	Revenue in billion (cf. 2014)	Notes
1	Anheuser-Busch InBev	$58.84 (43.6)	America / Belgium
2	Diageo plc	$32.38	UK
3	Heineken	$31.03 (22.8)	Netherland

Source: CompaniesMarketcap.com (2025), Created by the author

2.3.2.2 Tobacco

The tobacco industry began to decline after the release of the 1964 US Surgeon General's Report on Smoking and Health (Luther, 1964). According to WHO, smoking causes approximately four million deaths annually worldwide. This number is expected to rise to ten million by 2030, with 70% of those deaths occurring in developing countries. Despite widespread knowledge of its health risks and increasing regulation through taxation, nicotine addiction remains strong, making quitting difficult. Although sales and production volumes have declined, the industry continues to generate substantial profits. In 2014, global tobacco sales reached USD 528 billion. See below Table 2.3 top three companies in the world.

2.3.2.3 Gambling

Global casino revenue exceeded USD 150 billion in 2024, driven by the rapid growth of online gambling. While Las Vegas in the United States and Macau in China were once the centers of physical casino markets, deregulation across Asian countries and online platforms are becoming the new industry hubs. Online gambling is growing rapidly and is fundamentally transforming the industry's structure. The global online gambling market is expected to grow annually by approximately 8.7% through 2040, propelled by the proliferation of mobile devices, the adoption of cryptocurrencies, and AI-based personalization technologies, potentially surpassing traditional casino businesses in scale. See below Table 2.4 top three companies in the world.

2.3.2.4 Weapons: This Category Includes Cluster Bombs, Uranium-Based Weapons, Chemical and Biological Weapons, and Nuclear Arms

The global defense industry is dominated by the US corporations such as RTX (Raytheon) and Lockheed Martin. The aerospace and defense sector earns massive revenue annually, particularly from the US military—the world's most powerful armed force. The top 100 global military and defense contractors collectively generate more than USD 400 billion annually.

The Ottawa Treaty, which prohibits the use, stockpiling, production, and transfer of anti-personnel mines, has 162 signatories. However, key military powers such as the United States, Russia, and China have not signed the treaty. See below Table 2.5 top three companies in the world.

Table 2.3 Top three global tobacco companies (2024)

Rank	Company	Revenue in billion (cf. 2014)	Notes
1	China tobacco	$213	China
2	Imperial brands plc	$39.59	UK
3	Philip Morris international	$35.174 (26.7)	United States

Source: Emergen-research (2024), Created by the author

Table 2.4 Top three global gambling companies (2024)

Rank	Company	Revenue in Billion (cf. 2014)	Country
1	MGM resorts	$17.05 (9.5)	United States
2	Flutter entertainment	$13	Ireland/United States
3	Las Vegas Sands	$11.3 (11.4)	United States

Source: SNJ Today (2025), Created by the author

Table 2.5 Top three global weapons companies (2022)

Rank	Company	Revenue in Billion USD (cf. 2014)	Country
1	RTX corporation	$67.1	United states
2	Airbus	$66.6	EU
3	Lockheed Martin	$66.6 (35.4)	United States

Source: Wikipedia (2022), Created by the author

2.4 Statistical Comparison Methods

2.4.1 Z-Score Transformation of Stock Indices

In this chapter, various SRI and ESG index performances are statistically standardized using Z-score transformation. The Z-score indicates how far a given value deviates from the mean (in terms of standard deviations), allowing for meaningful comparisons across distributions—such as between different stock indices.

$$z = \frac{X - \mu}{\sigma}$$

This method enables fair and insightful comparisons between different investment strategies: traditional investing, negative screening, and positive screening. The statistical relevance of Z-transformation can be summarized as follows:

1. Ensuring Comparability: SRI and ESG indices are constructed from different investment universes, constituent stocks, and markets. Comparing raw index values may introduce bias. Z-score transformation allows for standardized comparison of how each strategy over- or under-performs relative to its mean.
2. Risk-Adjusted Evaluation: Z-transformation considers the spread of returns (standard deviation), enabling volatility-adjusted analysis. While SRI tends to

adopt a defensive (low-risk) profile, ESG strategies often reflect more active preferences, making Z-score transformation valuable in highlighting differences in return distribution.

3. Practical Application: Data scraping from online sources and computation can be done using Python in Google Colab. Alternatively, the following steps outline the Z-score process in Excel:

   A. Calculate the average:

      In a chosen cell (e.g., B1), enter:
      =AVERAGE(A2:A100).

   B. Calculate standard deviation (for the population):

      In another cell (e.g., B2), enter:
      =STDEV.P(A2:A100) (Note:. P for population & unbiased variance,. S for sample).

   C. Compute the Z-score:

      In cell B3 (next to your first data point), enter and autofill:
      =(A2 - B1)/B2.

2.4.1.1 MSCI World Index

The 2016 study identified the MSCI World Index as a widely used benchmark among institutional investors for global equity portfolios. The broader MSCI All Country World Index (ACWI) comprises 23 developed and 26 emerging markets and calculates market capitalization using free-float-adjusted values (Yahoo Finance 2026).

2.4.1.2 DJSI World Index

The Dow Jones Sustainability Index (DJSI), launched in 1999 by S&P Dow Jones Indices (S&P Dow Jones Indices (2026)) and Swiss sustainability ratings agency RobecoSAM (RobecoSAM 2025), is the world's first global ESG equity benchmark. It evaluates over 3500 leading global companies on "Economic," "Environmental," and "Social" dimensions, selecting top performers for inclusion.

2.4.1.3 FTSE4GOOD Index

The FTSE4Good Index was created in 2001 by the UK-based FTSE Group, originally founded by Pearson in 1995. This ESG benchmark evaluates companies based on environmental, social, and governance criteria. ESG scores are assessed according to industry- and region-specific standards, and high-risk sectors or companies violating international norms are excluded (FTSE4GOOD 2017a). The index is reviewed biannually with input from expert panels (Table 2.6).

Table 2.6 Z-Score correlation between MSCI World, DJSI, and FTSE4Good (FTSE4GOOD 2017b)

	Correlation
DJSI / FTSE	0.9690
MSCI / FTSE	**0.9743**
DJSI / MSCI	0.9355

Source: Author's calculation; DJSI and FTSE4Good (daily data: 1619 entries), MSCI (monthly data: 53 entries)

2.4.2 Text-Mining Analysis

To quantify ESG investment influence, we applied Latent Dirichlet Allocation (LDA) topic modeling to annual reports from three major German automakers: Volkswagen, Mercedes-Benz (Daimler), and BMW. The analysis covers 7 years (2017–2023), divided into pre-COVID (2017–2019) and post-COVID (2020–2023) periods, totaling 21 firm-years.

2.4.2.1 Interpretation of LDA Results[1]

Figure 2.3 displays Topic 1 to Topic 5 and their respective keyword distributions, characterizing each company's annual reports. The horizontal axis (Beta value) represents the probability of each keyword appearing within a given topic (Kobayashi,

[1] Python Code:

```
<ProgramCode ID=”PC1”>import os
import pandas as pd
import numpy as np
import matplotlib.pyplot as plt
import seaborn as sns
import re
import string
from sklearn.feature_extraction.text import CountVectorizer
from sklearn.decomposition import LatentDirichletAllocation
# ------------------------------
# 1. File Paths
# ------------------------------
file_paths_2017_2019 = [
    ---------
]
file_paths_2020_2023 = [
--------------
]
# ------------------------------
# 2. Preprocessing Functions
# ------------------------------
def preprocess(text):
    text = text.lower()
    text = re.sub(r'\d+', '', text)  # Remove numbers
    text = text.translate(str.maketrans('', '', string.punctuation))  # Remove punctuation
    text = re.sub(r'\s+', ' ', text)  # Normalize whitespace
    return text.strip()
```

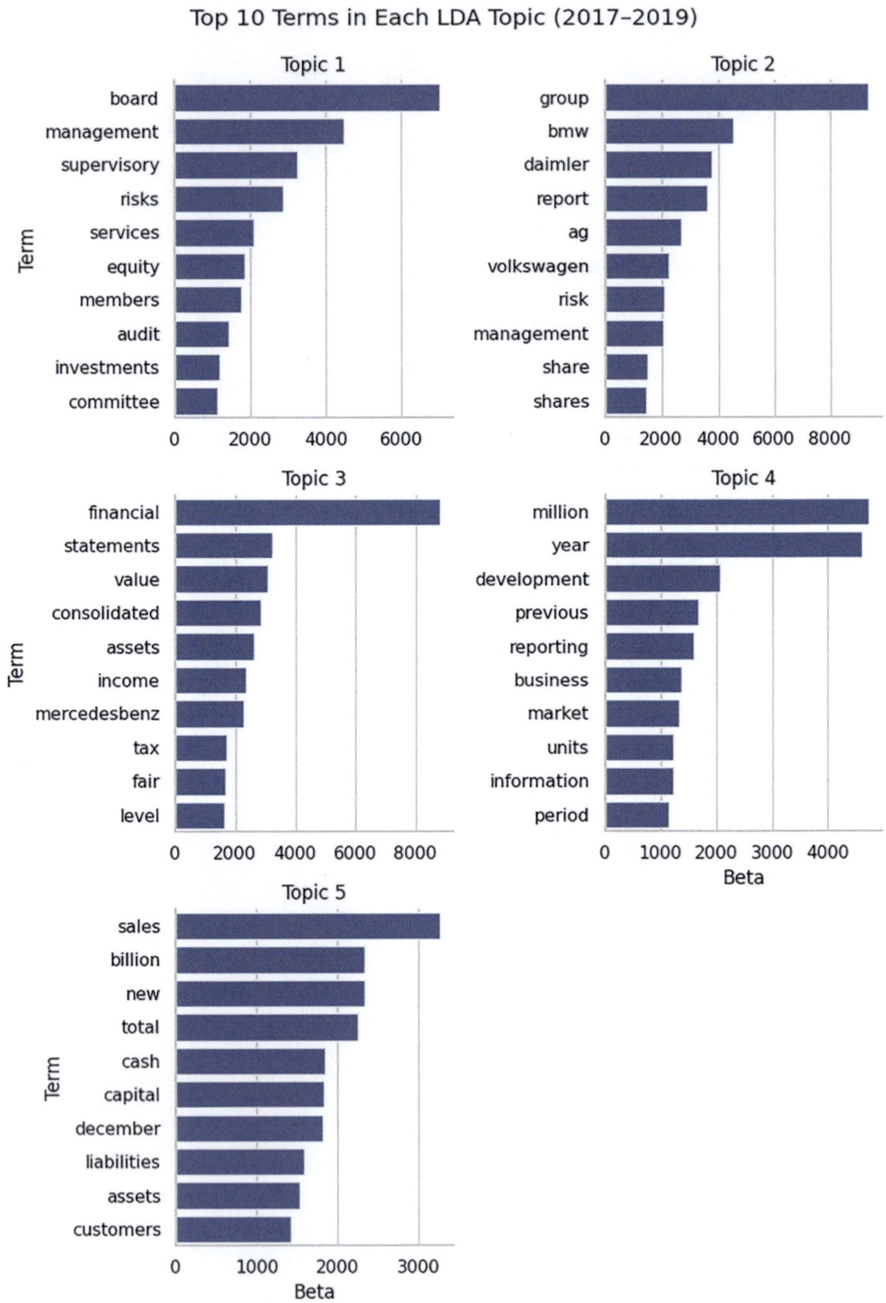

Fig. 2.3 LDA topics for 2017–2019 vs. 2020–2023. Source: Author’s calculation using R and Python

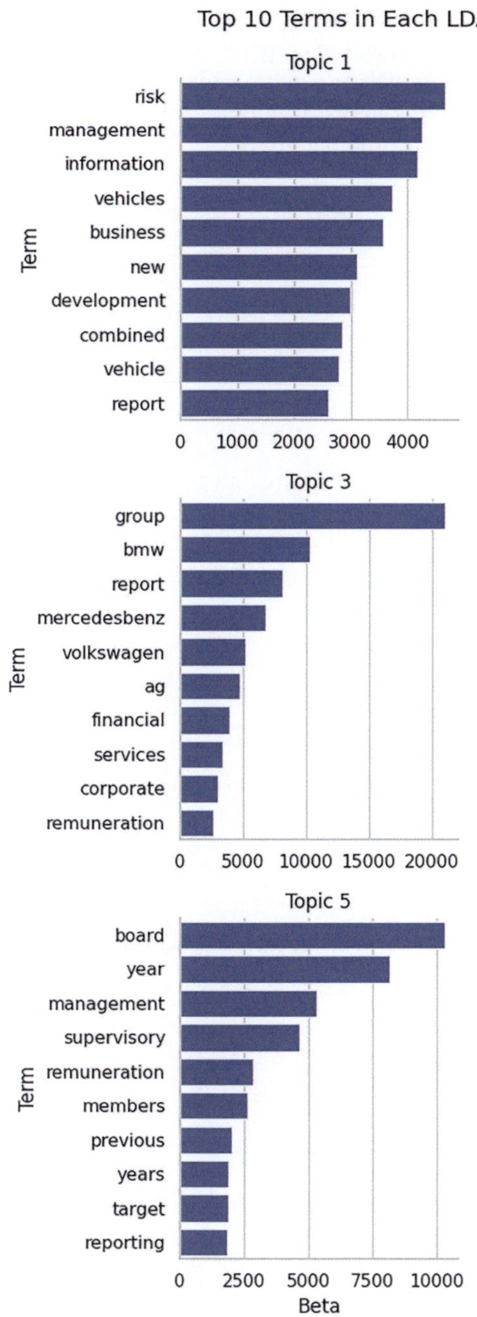

Fig. 2.3 (continued)

2023). For example, if "management" shows a high Beta value in Topic 1 or 2 (2017–2019), it suggests the topic is closely related to corporate governance.

Using Python and Google Colab (https://colab.research.google.com/) simplifies cloud-based LDA analysis. After installing the necessary libraries, topic modeling can identify vocabulary patterns and shifts in corporate discourse.

```
def load_texts(file_paths):
    all_texts = []
    for path in file_paths:
        with open(path, ‘r’, encoding=’utf-8’) as f:
            lines = f.readlines()
            processed_lines = [preprocess(line) for line in lines if line.strip()]
            all_texts.extend(processed_lines)
    return all_texts
# ------------------------------
# 3. Create DTM
# ------------------------------
def create_dtm(texts, max_df=0.99, min_df=2):
    vectorizer = CountVectorizer(
        stop_words=’english’,
        max_df=max_df,
        min_df=min_df
    )
    dtm = vectorizer.fit_transform(texts)
    feature_names = vectorizer.get_feature_names_out()
    non_empty_rows = np.asarray(dtm.sum(axis=1)).flatten() > 0
    dtm = dtm[non_empty_rows]
    texts_filtered = [text for i, text in enumerate(texts) if non_empty_rows[i]]
    return dtm, feature_names, texts_filtered
# ------------------------------
# 4. Build LDA Model
# ------------------------------
def run_lda(dtm, k=5, random_state=1234):
    lda = LatentDirichletAllocation(n_components=k, random_state=random_state)
    lda.fit(dtm)
    return lda
# ------------------------------
# 5. Visualize Topics
# ------------------------------
def plot_top_terms(lda_model, feature_names, top_n=10, title_suffix=""):
    topics = []
    for topic_idx, topic in enumerate(lda_model.components_):
        top_indices = topic.argsort()[-top_n:][::-1]
        for i in top_indices:
            topics.append((topic_idx + 1, feature_names[i], topic[i]))
    df_topics = pd.DataFrame(topics, columns=["Topic", "Term", "Beta"])
    g = sns.catplot(
        data=df_topics, kind="bar",
        x="Beta", y="Term", col="Topic", col_wrap=2,
        sharex=False, sharey=False, height=4
```

From 2017 to 2019, reports featured terms such as "board," "audit," "financial statements," and "sales," reflecting traditional governance and financial management priorities. In contrast, 2020–2023 reports increasingly used terms like "risk," "vehicles," "development," "emissions," "Porsche," and "remuneration," highlighting corporate focus on climate adaptation, innovation, and brand strategy.

Germany's 2020 environmental subsidy (Umweltbonus) significantly accelerated EV adoption. This is reflected in LDA outputs emphasizing "vehicles," "emissions," and "development," indicating how policy incentives influenced not only product strategy but also disclosure language and governance narratives. As subsidies were reduced after 2023, discourse shifted toward "risk," "legal," and "target," signaling rising structural uncertainty.

Notably, keywords like "management," "information," and "supervisory" suggest heightened focus on accountability, while words such as "combined," "group," and "corporate" indicate restructuring efforts and risk diversification strategies.

This illustrates how topic modeling helps visualize regulatory impacts on corporate language, enabling analysis of the intersection between energy policy and firm strategy.

2.4.2.2 EV Adoption Rates and Energy Mix

As shown in Fig. 2.4, the early 2020s saw major EV subsidy programs introduced in Europe, the United States, and Japan. In Germany, the 2020 Umweltbonus led to a surge in EV sales, mirrored in LDA analysis by increased mentions of "vehicles," "emissions," and "development" (JETRO 2020) This suggests that subsidies shaped not only innovation investments but also annual reporting language.

```
    )
    g.set_titles("Topic {col_name}")
    g.set_axis_labels("Beta", "Term")
    plt.suptitle(f"Top {top_n} Terms in Each LDA Topic ({title_suffix})", y=1.02)
    plt.tight_layout()
    plt.show()
    return df_topics
# -------------------------------
# 6. Execute and Save Analysis
# -------------------------------
# --- 2017–2019 ---
texts_17_19 = load_texts(file_paths_2017_2019)
dtm_17_19, features_17_19, _ = create_dtm(texts_17_19)
lda_17_19 = run_lda(dtm_17_19, k=5)
df_topics_17_19 = plot_top_terms(lda_17_19, features_17_19, title_suffix="2017–2019")
df_topics_17_19.to_csv("/content/lda_topics_2017_2019.csv", index=False)
# --- 2020–2023 ---
texts_20_23 = load_texts(file_paths_2020_2023)
dtm_20_23, features_20_23, _ = create_dtm(texts_20_23)
lda_20_23 = run_lda(dtm_20_23, k=5)
df_topics_20_23 = plot_top_terms(lda_20_23, features_20_23, title_suffix="2020–2023")
df_topics_20_23.to_csv("/content/lda_topics_2020_2023.csv", index=False)
```

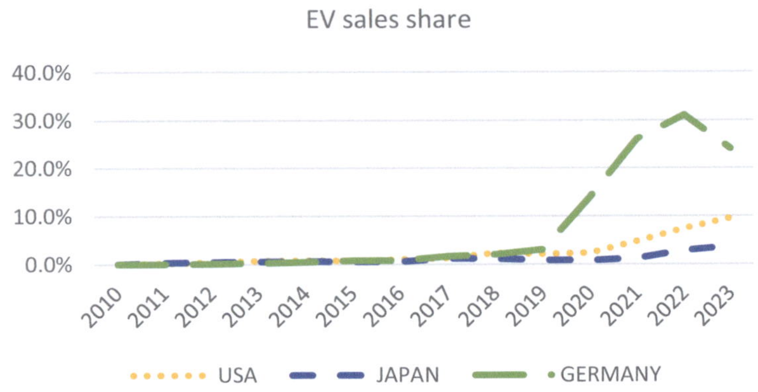

Fig. 2.4 EV share. Source: Author, based on [JETRO]

Germany's experience shows that subsidies can promote disruptive innovation and consumer behavior change. In the United States, similar effects are linked to Biden's green infrastructure policies (e.g., the Inflation Reduction Act). EV sales there are estimated at 40 million—double Germany's 24 million—even when adjusting for population size (IEA, 2024).

Japan's slower EV adoption reflects a shift in energy policy following the Fukushima disaster. With reduced reliance on nuclear energy, Japan emphasizes stable power supply and reexamines its energy mix. This complexity makes the alignment between EV adoption and power infrastructure a major policy challenge.

Furthermore, the structural impact of EVs on internal combustion engine (ICE) industries has raised employment concerns. Interviews in Germany revealed that some unions and suppliers see EV promotion as a threat to long-term business models. As a result, corporate reports increasingly emphasize governance and accountability.

LDA analysis revealed a growing emphasis on non-financial terms such as "CSR," "governance," and "disclosure." Figure 2.4 (EV Share) illustrates these shifts.

Japan's centralized power system—anchored in nuclear—has been transitioning toward decentralized, renewable sources since the 2011 Great East Japan Earthquake. In contrast, Germany and the United States had already started this transition earlier. Despite different approaches, each country must adapt energy policy to its unique regulatory infrastructure, including power generation, transmission, and distribution.

The shift to EVs is more than a decarbonization goal; it demands that automakers align product strategy with national energy conditions.

This is also related to the decentralized energy systems discussed below in connection with blockchain technology. Japan's energy system has undergone major changes: since the Great East Japan Earthquake in 2011, the use of nuclear power plants—previously the cornerstone and symbol of centralized power generation—has been restricted. As a result, Japan is now transitioning toward a renewed energy mix that incorporates renewable energy within a decentralized energy system.

By contrast, countries like the United States and Germany began transitioning to decentralized energy systems even before Japan's major shift triggered by the natural disaster. However, in all three countries, energy systems—being critical national infrastructure—must adapt to each country's specific regulatory frameworks governing electricity generation, transmission, and distribution.

In this context, the promotion of electric vehicles (EVs), which is the focus of this chapter, clearly serves as a driving force not only toward a decarbonized society but also demands that automobile manufacturers devise corporate strategies that align with each country's particular energy circumstances.

2.5 Conclusion

This chapter has illuminated the fundamental dilemma between Socially Responsible Investment (SRI) and ESG investing. The core tension lies between exclusion based on ethics (SRI) and inclusion based on accountability (ESG)—reflected in the methodological opposition between negative and positive screening.

The 1992 Cadbury Report's "Comply or Explain"principle marked a turning point, shifting investment focus from rigid exclusion toward transparency and investor choice—hallmarks of ESG investing.

However, the rise of ESG investing also brings the risk of greenwashing—masking externalities behind vague disclosures. SRI, while ethically clear, struggles to balance morality with economic efficiency. In the 2020s, environmental policies and EV subsidies have directly shaped corporate behavior, governance, and reporting content.

Emerging techniques like LDA-based text mining and blockchain-enabled real-time accountability offer a bridge between these two approaches. Ultimately, SRI and ESG should not be seen as mutually exclusive, but as complementary frameworks. Achieving a balance between ethics, sustainability, and market efficiency is the core challenge for future sustainable investing.

The chapter concludes that hybrid investment models—combining the values of SRI with the pragmatism of ESG—offer the key to a true paradigm shift from growth to coexistence.

2.6 Review Questions

1. Greenwashing and the Volkswagen Case.

   Explain the concept of "greenwashing" in the context of ESG investment. Using the Volkswagen "Dieselgate" case mentioned in the text, discuss how greenwashing can distort the evaluation of corporate environmental performance and affect market trust.
2. The Role of the Cadbury Report (1992).

Describe the significance of the Cadbury Report (1992) and its principle of "Comply or Explain."

How did this report change the framework of corporate governance and influence the evolution of ESG-oriented investment strategies?

3. Statistical Standardization (Z-Transformation).

   What is the purpose of using a Z-transformation when comparing different SRI and ESG stock indices? Explain how this statistical method helps ensure comparability and risk-adjusted evaluation among different investment strategies.

4. Text Mining and LDA Analysis.

   What is the purpose of applying LDA (Latent Dirichlet Allocation) analysis to corporate annual reports in ESG research? Based on the German automobile industry example, explain how topic modeling can reveal changes in corporate focus and governance over time.

5. The Decline of Sustainable Investment in the United States (2020–2022).

   According to the text, sustainable investment in the USA decreased between 2020 and 2022. Discuss the main reasons for this decline and explain how stricter reporting and anti-greenwashing standards influenced the total reported amount.

References

Barnett, M. L., & Salomon, R. M. (2002). Unpacking social responsibility: The curvilinear relationship between social and financial performance. *Academy of Management Review*, SIM:B1. Retrieved 2002, from https://scholar.google.com/citations?view_op=view_citation&hl=ja&user=27_fCmIAAAAJ&citation_for_view=27_fCmIAAAAJ:WF5omc3nYNoC

Blaine, T. (2020). From SRI to ESG:The origins of socially responsible and sustainable investing. *The Journal of Impact & ESG: Portfolio Management Research*. Retrieved from https://www.bailard.com/wp-content/uploads/2020/09/History-Socially-Responsible-Investing-and-ESG-Investing.pdf

Chang, E. C., Nelson, W., & Witte, D. (2012). Do green mutual funds perform well? *Management Research Review, 35*(8). https://doi.org/10.1108/01409171211247695

CompaniesMarketcap.com. (2025, July). *Top publicly traded alcoholic beverage companies by revenue*. Retrieved from, https://companiesmarketcap.com/alcoholic-beverages/largest-alcoholic-beverage-companies-by-revenue/?utm_source=chatgpt.com

Cowton, C. J. (2018). Socially responsible investing. *The Cambridge handbook of psychology and economic behaviour*. Retrieved from, https://www.cambridge.org/core/books/cambridge-handbook-of-psychology-and-economic-behaviour/socially-responsible-investing/5347A4233460B99C6430AEF45C3ACAB8

Crane, A., & Matten, D. (2010). *Business ethics, managing corporate citizenship and sustainability in the age of globalization* (4th ed.). OUP Oxford.

Emergen-research. (2024, May). *Top 10 companies in tobacco market in 2024*. Retrieved from https://www.emergenresearch.com/blog/top-10-companies-in-tobacco-market-in-2024?srsltid=AfmBOoqU3Y-4NZTzkXa_DbMceUPPRKV2J6_8xRAbv-6sMGMkvRzTNGO8&utm_source=chatgpt.com

Fortune Business Insights. (2025, June). *Alcoholic beverages market size, share & industry analysis*. Retrieved from https://www.fortunebusinessinsights.com/alcoholic-beverages-market-107439

FTSE4GOOD. (2017a). *Index inclusion rules for the FTSE4Good index series v1.9*. (The reference has been updated to the latest version (2025) and the access date has been added. Retrieved from https://www.lseg.com/content/dam/ftse-russell/en_us/documents/ground-rules/ftse4good-index-series-ground-rules.pdf

FTSE4GOOD. (2017b, August 29). *The historical index values of the FTSE4GOOD*. Retrieved from https://markets.ft.com/data/indices/tearsheet/historical?s=FT4GDBGL:FSI

Hirschman, A. O. (1970). *Exit, voice, and loyalty: Responses to decline in firms, organizations, and states*. Harvard University Press. Retrieved from http://www.hup.harvard.edu/catalog.php?isbn=9780674276604

IEA. (2024, July). *Global EV data explorer*. Retrieved from https://www.iea.org/data-and-statistics/data-tools/global-ev-data-explorer

JETRO (2020) Subsidy amounts for the purchase of low-emission vehicles temporarily increased. Japan External Trade Organization, 15 July 2020. Available at: https://www.jetro.go.jp/biznews/2020/07/fa522ac027b15478.html

Knoll, M. S. (2002). *Ethical screening in modern financial markets: The conflicting claims underlying socially responsible investment*. Business Lawyer, 57 Bus. Law. 681. Retrieved August 29, 2017, from https://www.law.upenn.edu/fac/mknoll/ethicalscreening.pdf

KobayashiYuichiro. (2023). Easy text analytics with R. (VersionJapanese, 訳) Ohmsha, Ltd.

Luther, T. L. (1964). *The first report on smoking and health*. The Surgeon General's Advisory Committee. Retrieved from https://www.cdc.gov/tobacco-surgeon-general-reports/about/history.html?utm_source=chatgpt.com

Mans-Kemp, N. (2014). *Corporate governance and the financial performance of selected Nadia Johannesburg Stock Exchange Industries*. Dissertation presented for the degree of Doctor of Philosophy in Business Management in the Faculty of Economic and Management Sciences at Stellenbosch University. Retrieved August 29, 2017, from https://scholar.sun.ac.za/bitstream/handle/10019.1/95957/manskemp_corporate_2014.pdf?sequence=2

Marcos, C., & Vanessa, S. (2024). *How blockchain provides confidence in ESG reporting to accelerate sustainability transformation (SX)*. Fujitsu. Retrieved from, https://activate.fujitsu/-/media/Project/Fujitsu/Fujitsu-Activate/uvance/digital-shifts/How-blockchain-provides-confidence-in-ESG-reporting-to-accelerate-sustainability-transformation.pdf

Matthews, C. (2016). *The 5 biggest vice industries in the world*. Retrieved from http://fortune.com/2016/03/24/vice-industries/

Yahoo Finance (2026) URTH (MSCI World-related instrument) historical price data. Yahoo Finance. Historical price data were retrieved programmatically from Yahoo Finance using Python (e.g., the yfinance package) based on the ticker "URTH". Available at: https://finance.yahoo.com/quote/URTH/

Norges Bank. (2020). *The list of companies that are excluded from the Government Pensions Fund Global*. Norway. Retrieved from https://www.nbim.no/en/the-fund/responsible-investment/exclusion-of-companies/

Parkin, S; Sommer, F; Uren, S. (2003). Sustainable development: Understanding the concept and practical challenge. *Proceedings of the Institution of Civil Engineers Engineering Sustainability*.

Piketty, T. (2014). *Capital in the twenty-first century*. Belknap Press World. Retrieved from https://dowbor.org/blog/wp-content/uploads/2014/06/14Thomas-Piketty.pdf

Poelloe, A. (2010). *Is there a trade-off between social responsibility and financial performance?* Masters Thesis. Erasmus University Rotterdam. Retrieved August 29, 2017, from https://thesis.eur.nl/pub/8254/Poelloe,%20A.%20287713%20id%20thesis8254%20.pdf

RobecoSAM. (2025). *RobecoSAM Exclusion Policy December 2025*. Retrieved from https://www.robeco.com/files/docm/docu-exclusion-policy.pdf

Scott, C. M. (2009). *Green economics*. Earthscan Publishing. Retrieved from http://www.gci.org.uk/Documents/128075741-Green-Economics-an-Introduction-to-Theory-Policy-and-Practice.pdf

SNJ Today. (2025, July). *Largest gambling companies by revenue in 2024.* Retrieved from https://snjtoday.com/largest-gambling-companies/?utm_source=chatgpt.com

The Committee on the Financial Aspects of Corporate Governance. (1992). *Report of the committee on the financial aspects of corporate governance.* London Stock Exchange. Retrieved June 1, 2018, from http://www.ecgi.org/codes/documents/cadbury.pdf

The Domini Index. (2026). Retrieved from http://www.domini.com

S&P Dow Jones Indices (2026) The historical index values of the Dow Jones Best-in-Class Indices (formerly known as the Dow Jones Sustainability Indices, DJSI). Available at: https://www.marketplace.spglobal.com/en/solutions/dow-jones-best-in-class-indices-(dfb058e9-fd4f-4f0d-b325-8aa18d7a0373)

The Global Sustainable Investment Alliance. (2025). *The global sustainable investment review 2022.* Retrieved from https://www.gsi-alliance.org/members-resources/gsir2022/

The Global Sustainable Investment Alliance (GSIA). (2016). *The global sustainable investment review 2016.* Retrieved August 29, 2017, from http://www.gsi-alliance.org/wp-content/uploads/2017/03/GSIR_Review2016.F.pdf

The Global Sustainable Investment Alliance (GSIA). (2022). *The global sustainable investment review 2022.* Retrieved August 29, 2022, from https://www.gsi-alliance.org/members-resources/gsir2022/

Wikipedia. (2022). *Aerospace manufacturer.* Retrieved from https://en.wikipedia.org/wiki/Aerospace_manufacturer?utm_source=chatgpt.com

Yulia, S. (2017). Evolution of responsible investment. *Principal of responsible investment.* Retrieved June 1, 2018, from https://fs.moex.com/files/15304

Chapter 3
"Pricing Game" for tacit collusion and Passive Investment

Abstract This paper aims to figure out the structural factors of tacit collusion from the perspective of the oligopolistic market. A two-step approach is adopted to analyze this phenomenon. As pricing mechanisms shift from traditional methods to computational algorithms, herein termed the "pricing game," new forms of collusion are expected to emerge. First, game theory is applied toward an understanding of this unspoken collusion, which involves interaction between different parties. A potential new form of collusion is identified as having been created by information signals in the price networks. Second, firms are owned by overlapping sets of investors (passive investors), and their incentives to compete are thereby reduced. Investors are rapidly shifting their investment allocations from active to passive management (ETF; Exchange-Traded Funds), in response to the complexity of asset management and the excess liquidity from central banks around the industrial world. This trend has accelerated during the last decade. A potential solution for this situation may be found in family ownership, as a countervailing power for healthy competition.

Keywords Tacit collusion · Oligopoly · Pricing algorithm · Game theory · Passive and active investors

3.1 Introduction

"Googling," "Hotel and Flight Booking," and other modes of accessing information are now commonplace in our everyday lives. They have been described as "algorithms-as-institutions" (Robyn & Danah, 2018), raising questions about the extent to which human decision-making processes will ultimately be replaced, in certain cases, by machines (OECD, 2017a). However, there are a growing number of companies using a computer algorithm to improve their pricing models, tailored

K. Shimizu, *ESG Investment, German Industrie 4.0, and Blockchain*,
https://doi.org/10.1007/978-981-95-6927-4_3

services and to predict market trends, not simply for customers but, particularly, to maximize profit. The pricing model is not only constructed by traditional supply and demand, but also by using an ever faster, cheaper, and more powerful computer, which can fully automate processes to optimize a particular interaction between parties.

This study discusses a price which shows their product value, cost structure, and their competitiveness. For example; Flight ticket prices are approximately 3–7% higher in the American airline route than would be the case under special condition of ownership (Azar et al., 2017). Also, many fees for banking services are recorded at historical highs in America despite offering the lowest interest rate in recent time (Jose et al., 2016). For example, the price of eating or drinking at an amusement park or high prestige restaurant is higher than downtown, of course, because the price is affected by location (i.e. a special condition). Galbraith, K. notes the dependence effect. Demand for goods and services is organic if the consumer suffers a privation. However, new demands are created by advertisers who benefit from increased consumer spending and to provide a less urgent supply. The goal is to fight against those affirmations based on "conventional wisdom" (Galbraith, 2010), the so-called "institution."

Generally, firms compete with one another by offering the best prices in a particular market. The "pricing game" can lead to cartels. *Industries using sophisticated computer algorithms can promptly calculate interactions with competitors and update their prices while also considering their next move* (Ezrachi & Stucke, 2015). This computer algorithm may represent a signal for the counterpart; however, it tends to remain undetected, most notably by antitrust authorities.

This research aimed to identify the structural factors of tacit collusion, which is when two firms agree to play to a particular condition without explicitly stating an exact price. For example, Company A holds more than 50% of the market share and sets the price at 100. There is then the question of how counterpart Company B should behave under the condition (Figs. 3.1 and 3.2). If Company A is a price leader, Company B usually follows it. Both companies depend on the action of their competitor, here Company B. When Company B reduces the price to 80, it can gain market share. The normal reaction of Company A would be to reduce its price to 80 also. The overall response is good for the consumer. However, there is a way to avoid reducing their price yet keep their profit; this is called "tacit collusion."

The price mechanism is shifting from the traditional method to a computational algorithm. New collisions can form via computational signals. Also, in general, shareholders are becoming decentralised and institutionalized. A shareholder's money is in trust with institutional investors who are interested only in high and stable dividends and dislike issues that could reduce these dividends. Furthermore, a money manager can create a smart investment approach, such as an exchange-traded fund (ETF). Holders of ETFs are shareholders who desire high and stable dividends but are not keen to be directly involved with company management. Galbraith suggests a countervailing power, which optimizes stakeholders under certain conditions. Family ownership can provide such countervailing power because it imposes identity onto a company's culture and strategy.

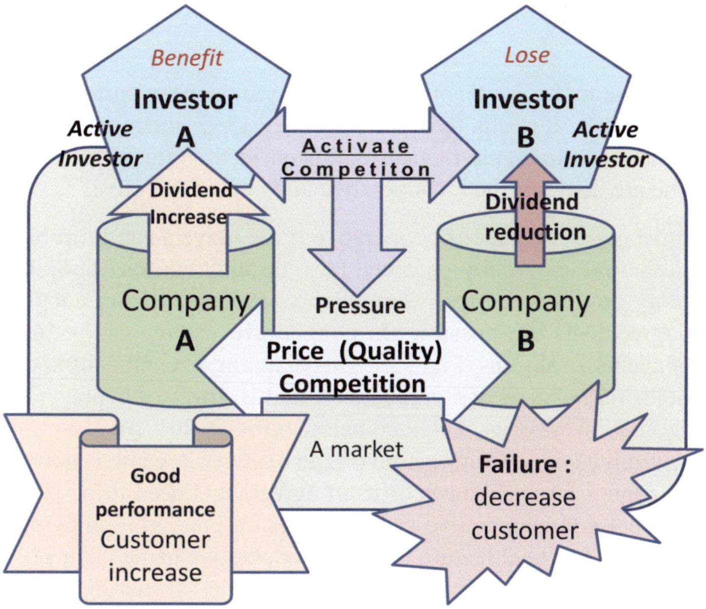

Fig. 3.1 A competitor that is owned by different investors

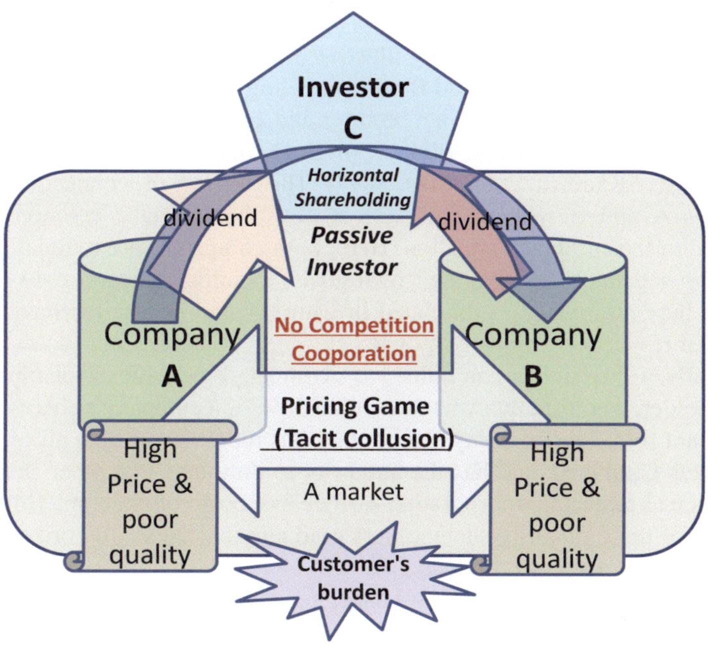

Fig. 3.2 A competitor that is owned by same investor

3.2 Game Theory

The "game" in game theory denotes the interaction between parties that are acting in their own interests. It could, therefore, also be called "interaction theory," and is applicable to the phenomenon of tacit collusion in the oligopolistic market. Two types of game are distinguished: cooperative and non-cooperative.

- *Cooperative game*; A game is cooperative if the players can form binding commitments that are externally enforced (e.g. through contract law). Cooperative game theory model scenarios, when agents can benefit from cooperating, and binding agreements are possible. In cooperative games, actions are taken by groups of agents, coalitions (Fig. 3.2). This diagram schematically shows investor C who holds both shares of Companies A and B. Investor C player plays a role as a horizontal investor, as explained below (OECD, 2017b).
- *Non-cooperative game* (Competition on the market); A game is non-cooperative if players cannot form alliances or if all agreements need to be self-enforcing. The traditional non-cooperative game focuses on predicting individual player's actions. Also, the critical feature of a non-cooperative game is the absence of an external authority to enforce rules. For example, a sharing of information between Competitors A and B is not permitted under the Sherman Act and the US federal antitrust law (Department of Justice, 2026) or "Gesetz gegen Wettbewerbsbeschränkungen (GWB) [Bundesministerium der Justiz und für Verbraucherschutz, 2026]" and "Japanese Act on Prohibition of Private Monopolization and Maintenance of Fair Trade [Japanese Law Translation, 2026]."

Figures 3.1 and 3.2 schematically illustrate a common non-cooperative situation [Mizuta, 2018]. Companies A and B are competing with each other. There are several types of competition, which act both for and against each other, such as a joint venture, co-operation, strategic alliance, original equipment manufacturer (OEM), and Japanese "Keiretsu." (Shimizu 2017) The degree of connection between Companies A and B will crucially affect how their specific resources will be exchanged in the combined situation. Here, both companies are rational economic agents who separately achieve their own goals. Usually, companies are owned by investors, here designated as unrelated investors A and B. Their behavior will be that of their respective Company A or B.

Generally, a shareholder can influence a company's activities. Notably, a majority shareholder, who usually owns more than 50% of a company's shares, can play an important role. This majority shareholder has more power than all of the other shareholders combined, and has the authority to do things that other shareholders cannot, such as replacing a corporation's officers or board of directors. However, the attribution of these majority shareholders gradually changes into horizontal shareholding, which is discussed in Sect. 2.1.

In the context of corporate governance, separation of ownership and management means that the management of the company can authorize its activities under dispersed ownership; no one shareholder has enough shares to be able to control the company. There is also the stakeholder's view of a firm (Steinmann et al, 2013). In defining “Stakeholder Theory“Freeman et al. (2010) states: “The firm is a system of stakeholder management within the larger system of the host society that provides the necessary legal and market infrastructure for the firm’s activities” (Freeman et al., 2010). Business is about how customers, suppliers, employees, financiers (e.g. stockholders, bondholders, banks), communities and managers interact and create value. Therefore, “Stakeholder Theory” and a cyber-platform can deal fairly well with “Game Theory.”

Figure 3.1 shows a competitor that is owned by **different** investors. These are investors who are actively involved in formulating strategy for Companies A and B, so-called active investors. Figure 3.2 shows a company that is owned by the **same** investor. The investor is not involved in strategy formulation, and is thus called a passive investor (here, investor C).

Figures 3.1 and 3.2 illustrate the analysis for maximum profit (dividend) for investors A and B, and investor C, in the two scenarios. The active investors in Fig. 3.1 are expected to promote competition between Companies A and B and within the market at large. In contrast, investor C (in Fig. 3.2) reduces risk through passive investment, in comparison with the reduction of dividends by competition between Companies A and B. Investor C, therefore, creates a stable profit condition. For example, ETFs track stock indexes. A stock index is computed from the prices of selected stocks (typically a weighted average). The selected stocks are chosen according to many factors such as a market-cap, free-float, and other portfolio-related criteria.

- *A passive investor* (investor C in Fig. 3.2) invests for the long haul. Passive investors are based on a “random walk,” which is a statistical phenomenon. The passive investor believes a variable follows no discernible trend and moves seemingly at random. Malkiel’s random walk theory points out that the price of securities moves randomly, therefore, it is impossible to predict future price movement, either through fundamental or technical analyses. One type of passive-investor style involves a buy-and-hold mentality (Malkiel, 2019).
- *An active investor* takes a hands-on approach and requires that someone act as portfolio manager. The goal of active money management is to beat the stock market’s average returns such as that of the ETF or other leading indexes. A hedge fund uses this investment style. A portfolio manager usually supervises analysts who examine qualitative and quantitative factors.

3.2.1 Horizontal Shareholdings

Horizontal shareholdings have overlapping ownership by financial investors. It exists when a standard set of investors own significant shares in corporations that are horizontal competitors. Institutional investors have become leading shareholders in the airline, banking, computing, and pharmacy markets. Several empirical studies have confirmed that these horizontal shareholdings have anticompetitively affected airline and banking markets (Einer, 2017).

Institutional investors are specialized financial institutions that invest capital from a large number of investors on their behalf, pursuing the goal of maximizing returns at a reasonable risk. Institutional investors include insurance companies, asset managers, investment and pension funds, as well as banks and sovereign wealth funds. The German Monopolekommission focuses on investigating equities and differentiating between active and passive investment strategies. Active investors may choose between value, growth, and combination strategies (e.g., growth at a reasonable price (GARP)).

Passive investors typically engage in index-based diversified investment through exchange-traded funds (ETFs). In the case of the German stock index DAX40, the free-float ratio—shares actually available for trading on the market—has remained around 50–60%.

Investors who hold these free-float shares tend to place greater emphasis on the risk-reduction benefits of diversification than on the exercise of voting rights. Their interests therefore differ from those of active investors such as family-owned firms, foundations, or strategic shareholders who are more directly involved in corporate management. This distinction is later discussed in this chapter as a form of "countervailing power," functioning as a market-based constraint on corporate behavior.

According to the DIRK report, the top ten investors in the DAX30/40 index collectively held approximately 33.3% of all shares in 2021, with major institutional investors such as Vanguard, BlackRock, and Norges Bank dominating these holdings. As a result, large passive asset managers collectively hold a substantial portion of the free-float, creating new challenges for corporate governance. In particular, the concentration of voting rights among these investors may generate upward pressure on stock prices through the mechanism known as "Horizontal shareholdings" in which multiple firms are owned by the same diversified shareholders. (DIRK, IPREO, 2018) (DIRK (Deutscher Investor Relations Verband), 2024).

3.3 The Several Signals (Artificial Intelligence, Machine Learning, and Deep Learning)

Regarding the use of computer techniques below will create more secret collusion, and as a result, manipulates a price setting (Maurice & Ariel 2016). The stronger, more powerful algorithms will likely prevail and dominate the market. A simple scenario, we use computers to execute their instructions directly. Firms are owned by overlapping sets of investors (passive investors), and their incentives to compete are thereby reduced, explained about 2.1 Horizontal shareholding. The computer techniques are developing step by step, see 1–3 below.

1. Artificial Intelligence (AI).

   Chaotic data (now referred to as "big data") are generated through our daily experiences. Norbert Wiener first suggested that chaotic data could be regulated and organised using cybernetics as an approach to controlling the chaos (Norbert,

1950) and he is now considered to have been a pioneer in artificial intelligence (AI). The signals involve the following concepts: artificial intelligence (1950s-), machine learning (1980s-), and deep learning (2010s), in chronological order.

2. Machine learning.

   The machine is "trained" using large amounts of data and algorithms that enable it to learn how to perform a task. There are several types of learning pattern: 1. Supervised learning: the computer learns a general rule. 2. Unsupervised learning: no labels are given to the learning algorithm so that the hidden structures or patterns in the data may be discovered. 3. Active learning: the computer can only obtain training labels. 4. Reinforcement learning: feedback is given to the program's actions in a dynamic environment, such as a self-driving vehicle.

3. Deep learning. (Nvidia 2026)

   As you know, Go is a strategy board game for two players. The game was invented in China more than 2500 years ago. Alphabet Inc. (Google DeepMind) developed AlphaGo which is a computer program for playing the game of Go. AlphaGo has beaten a human professional Go player, who is the world no. 1 ranked player, Ke Jie in 2017. AlphaGo's algorithm uses a combination of machine learning.

3.3.1 Pricing Game

An agreement may be established as humans collude through the medium of computers (Table 3.1 Computer collusion scenario). The computer algorithms serve as the messengers in the sense that the cartel members program the computers to help the cartel. Table 3.1 shows three types of collusion: "messenger," "hub-and-spoke," and "autonomous machine." However, these are only recognized behind closed doors (tacitly).

- *Messenger:* This type of collusion occurs when a competitor agrees to use the same pricing algorithm under the same conditions, even if there is no agreement on direct price. As with the agreement to use the formula, the cartel is also established under the traditional cartel concept.
- *Hub and Spoke:* If a competitor happens to use the same pricing algorithm provided by a third party, it does not immediately cause a problem with respect to antitrust law. However, in the case where competitor price data themselves are

Table 3.1 Computer collusion scenario (Ezrachi & Stucke, 2015 p. 1784)

Model of collusion	Usage/Type/Characteristics of	Evidence of agreement
Messenger	Competitive companies use the same pricing algorithm	Strong evidence of horizontal agreement
Hub and spoke	Machine learning	Likely effect
Autonomous machine	Deep learning	Unclear

shared through third parties and they are mutually recognized, a hub-and-spoke type of cartel may result (Obermaier 2016).

- *Autonomous Machine*: When autonomous pricing algorithms make price adjustments irrespective of the intention of the operator, there is no artificial interaction between competitors. The cartel concept can be difficult to apply in such a situation.

3.4 Assets Under Management

"Horizontal Shareholding,"mentioned above in Sect. 3.2.1 based on respect to passive funds. Horizontal shareholding means that funds investing in shares will be majority shareholders of many companies in one industry. Usually, investors will apply pressure on corporate managers to increase the competitiveness of the holding company, which will win market share and the corporate value will rise. This will drive up the stock price and improve profit.

However, the situation will be different for the investor who holds shares of many companies in the same industry. A highly competitive market will lead to price reductions and, thereby, a loss in value of the company. There is a possibility that the total profits of both companies could actually be higher in such a situation, not because of price competition but, rather, because of "tacit collusion" in a low-growth market.

The shareholding structure may support such tacit collusion. Global assets under management are expected to rise from US$84.9 trillion in 2016 to US$145.4 trillion in 2025 according to PricewaterhouseCoopers (PwC). This firm predicts a compound growth rate of almost 6%. Figure 3.3 shows the share of active, alternative, and passive global assets under management. Traditional active management will

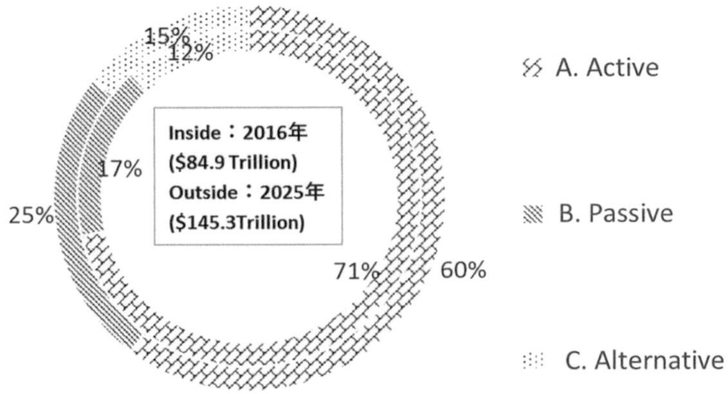

Fig. 3.3 The active, alternative, and passive shares within global assets under management (PwC, 2017)

continue to be the core of the industry in 2025. However, it will grow at a less rapid pace than the passive and alternative strategies.

PwC forecasts that active management will decrease from 71% in 2016 to 60% in 2025. Passive investment will gain significant market share, rising *from 17% to 25%*, while alternatives will increase from 12% to 15% over the same period. Passive investment will more than double, from US$14.2 trillion to US$36.6 trillion because of cheap passive ETFs. Alternatives will increase from US$10.1 trillion to US$21.1 trillion.

3.5 Conclusion and Solution

This study presents a two-step approach to examining the structure of tacit collusion.

First, we show that pricing mechanisms have gradually shifted from traditional methods toward new forms of "price games." Computer algorithms can now influence price increases—both for financial assets and goods and services—thereby creating new forms of collusion, including horizontal shareholding. These mechanisms are increasingly discussed as potential contributors to global inflation.

Second, the ownership of many corporations is now dominated by passive investors, especially the "Big Three" institutional investors: BlackRock, Vanguard, and State Street. Investors have shifted their funds into ETFs because they offer low costs, flexible trading, and effective diversification. Since the 2008 financial crisis, ETFs have also been used by central banks as tools to provide market liquidity.

Despite their convenience, ETFs may accelerate the hollowing out of voting rights in corporate governance. Corporate governance, defined as the relationship among stakeholders, managers, and the board of directors, is crucial for firm oversight. However, the Big Three face a shortage of personnel capable of exercising the vast voting rights associated with their highly diversified portfolios (DIRK (Deutscher Investor Relations Verband), 2024).

In Germany, these major index-tracking investors hold more than 5% of the institutional free-float and own a significant share of companies listed on the DAX index (DIRK, IPREO, 2018). (Azar et al., 2017).

Family ownership offers a potential counterbalance to these developments. Family shareholders tend to prioritize governance over short-term profit maximization and can therefore provide healthy competitive tension. Historically, family ownership in DAX30 companies accounted for 19.01%, compared with 16.58% held by foreign investors (Monopolkommission 2018).

Figure 3.4 classifies identifiable shareholders holding more than 1% of shares into six categories and illustrates the historical ownership structure of German listed companies from 1982 to 2022:

1. Free-float (publicly tradable shares)
2. Family ownership (individuals or family foundations)
3. Germany's Top 100 companies

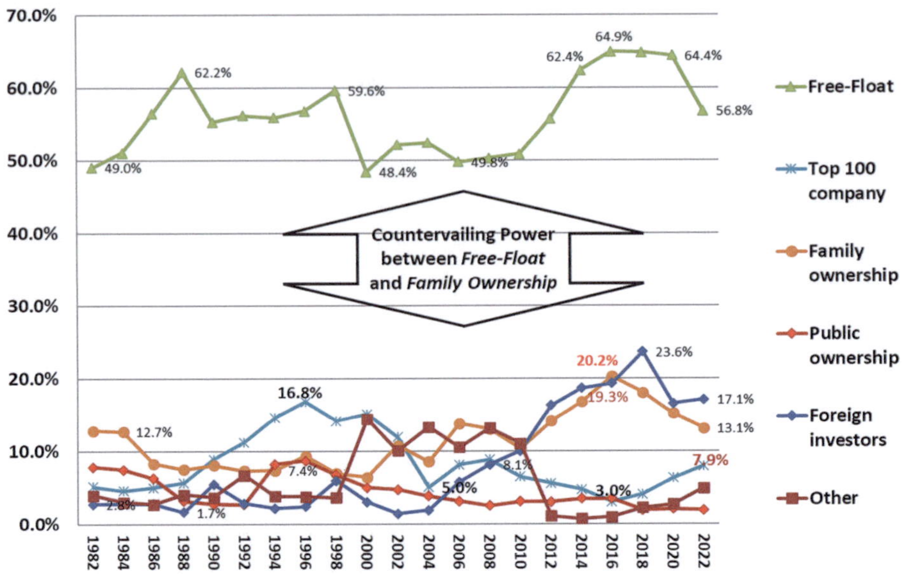

Fig. 3.4 Historical shareholding structure of the DAX30

4. Public-sector ownership
5. Foreign investors
6. Other

1. *Free-float*:

   Free-float refers to shares freely traded on the market and has remained between 50% and 60% over several decades. German listing regulations require companies to maintain at least 25% free-float. Locked-in shares held by the Top 100 firms, families, individuals, or the government are excluded from this category.

   As of 2022, most of the free-float (56.8%) is held by institutional investors, while family ownership (13.1%) provides a potential counterbalance. Given the expected further expansion of passive investment, this relationship is becoming increasingly important.
2. *Family ownership:*

   This category includes ownership by individuals or family foundations. Family ownership typically involves a direct or indirect majority stake and some level of influence over corporate strategy (Prof. Dr. Kirchdörfer, 2011).

   Family ownership (13.1%) is roughly comparable to the share held by foreign investors (17.1%). Since the 2000s, globalization and the employment-generating role of family firms have made this ownership form more attractive to policymakers.

3. *German Top 100* company:

   These firms contribute significantly to economic output and employment. Although their ownership share declined for many years, recent data show an increase from 3.0% to 7.9%. This change reflects shifts in family holdings, strategies to avoid golden-share regulations in the EU—such as Porsche's ownership of Volkswagen—and divestments by Siemens subsidiaries to increase free-float.
4. *Public ownership*;

   Public-sector ownership has decreased markedly over the past two decades, yet it remains important from a socio-political perspective. Government representatives often serve on the boards of companies in which the state holds shares.

   Privatization of major utilities, such as RWE AG, has reduced the weight of this category. Nevertheless, the public sector continues to provide a balancing framework between free-float and family ownership.
5. *Foreign investors*:

   The growth of institutional investors—such as global pension funds and investment trusts—has led to the widespread use of index-based investment strategies. In fact, institutional investors are estimated to hold about 40% of the world's total stock market capitalization. As overseas funds have come to own large volumes of shares, monitoring of corporate management and the frequency of shareholder proposals have increased.
6. *Other*:

   Shares of companies outside the German Top 100 companies and institutional investors in Germany as well as capital shares of co-operatives and unidentifiable participations are included in this category.

Family firms also play a significant role in employment.

Between 2006 and 2014, domestic employment in Germany's 500 largest family-owned firms increased from 2.67 million to 3.17 million—an increase of nearly 19%. In contrast, the 27 DAX-listed companies increased their employment by only 1.5% during the same period (Dr. Gottschalk & Egeln, 2017).

Furthermore, branding strategies in family businesses have become an important research topic. These firms often develop integrated structures—both vertically and horizontally—not only through physical production but also through the carefully cultivated symbolic and brand image associated with the company.

3.6 Review Questions

1. How are pricing mechanisms evolving as algorithmic decision-making becomes increasingly integrated into markets?
2. In what ways are advanced forms of tacit collusion emerging through algorithmic pricing and horizontal shareholding?
3. To what extent can modern capital market structures be considered drivers of global inflation?

4. How does the growing dominance of passive investors reshape corporate ownership and control?
5. Does the expansion of ETFs contribute to a hollowing out of corporate governance, and if so, how?
6. What problems arise from the mismatch between passive ownership and the need for active monitoring?
7. In what ways do family firms function as a countervailing force in contemporary capital markets?
8. Why is diversity in ownership structures—such as free-float, family ownership, public-sector holdings, and foreign investors—important for corporate governance?
9. How are foreign institutional investors influencing corporate behavior and strategic decision-making?
10. What social and economic contributions do family-owned enterprises make, and why are these contributions important?

References

Azar, J., Schmalz, M., & Tecu, I. (2017). Anti-competitive effects of common ownership. *FTC microeconomics conference 2015* [Online] 2017. https://www.ftc.gov/system/files/documents/public_events/633601/azar_presentation.pdf

Bundesministerium der Justiz und für Verbraucherschutz. (2026). Act against Restraints of Competition (Competition Act—GWB). [Online] [Cited: 2 26, 2026.] http://www.gesetze-im-internet.de/englisch_gwb/.

Department of Justice. (2026). Antitrust laws and you [Online] [Cited: 2 26, 2026.] https://www.justice.gov/atr/antitrust-laws-and-you.

DIRK (Deutscher Investor Relations Verband). (2024). Who owns the German DAX? *The Ownership structure of the German DAX in 2023*. [Online] 2024. https://www.dirk.org/wp-content/uploads/2024/06/DAX-Studie-Wem-gehoert-die-Deutschland-AG-11.0.pdf.

DIRK, IPREO. (2018). Investoren der Deutschland AG 5.0. [Online] 6 2018. https://www.dirk.org/dirk_webseite/static/uploads/Investoren_der_Deutschland_AG_5-0_DAX-Studie-2018_Ipreo_DIRK.pdf.

Dr. Gottschalk, S., Egeln, J., & Kinne, J. (2017). Die volkswirtschaftliche Bedeutung der Familienunternehmen. [Online] 2017. http://ftp.zew.de/pub/zew-docs/gutachten/2017_Studie_VolkswirtschaftlicheBedeutungFamilienunternehmen.pdf.

Einer, E.. (2017). The growing problem of horizontal shareholding. [Online] 2017. https://www.competitionpolicyinternational.com/the-growing-problem-of-horizontal-shareholding/.

Ezrachi, A., & Stucke, M. E. (2015). Artificial Intelligence & Collusion: When Computers Inhibit Competition. [Online] 2015. https://illinoislawreview.org/wp-content/uploads/2017/10/Ezrachi-Stucke.pdf.

Freeman, E., et al. (2010). *Stakeholder theory*. Cambridge University Press.

Galbraith, J. K. (2010). *The affluent society and other writings 1952–1967*. Literary Classics of the United States.

Steinmann, H., Schreyögg, G., & Koch, J. (2013). *Management: Grundlagen der Unternehmensführung Konzepte—Funktionene—Fallstudien Gebundene* (Auflage: 7 ed., p. 47). Springer Gabler.

Japanese Law Translation. (2026). Act on prohibition of private monopolization and maintenance of fair trade [Online]. Cited February 26, 2026, from https://www.japaneselawtranslation.go.jp/en/laws/view/4767.

Jose, A., Sahil, R., & Martin, S. (2016). *Ultimate ownership and bank competition*. SSRN.

Malkiel, B. G. (2019). *A random walk down Wall street: The time-tested strategy for successful investing*. W W Norton & Co Inc.

Maurice, S. E., & Ariel, E. (2016). How pricing bots could form cartels and make things more expensive. *Harvard Business Review* [Online]. October 27, 2016, from https://hbr.org/2016/10/how-pricing-bots-could-form-cartels-and-make-things-more-expensive.

Nvidia (2026). Deep Learning *NVIDIA* [Online]. https://developer.nvidia.com/deep-learning

Mizuta Takanobu. (2018). Sparks Special report. Online, April. 23 2018. http://www.mizutatakanobu.com/201806.pdf.

Monopolkommission. (2018). *Wettbewerb 2018*. Nomos-Verlag, 2018. S. 90,91. Bd. XXII. Hauptgutachten der Monopolkommission.

Norbert, W. (1950). *The human use of human beings; Cybernetics and society* [Online]. http://21stcenturywiener.org/wp-content/uploads/2013/11/The-Human-Use-of-Human-Beings-by-N.-Wiener.pdf/2014.05.10.

Obermaier, R. (2016). *Industrie 4.0 als unternehmerische Gestaltungsaufgabe* (p. 18-19). SpringerGabler.

OECD. (2017a). *Algorithms and collusion*. 2017. p.14 Box 5. Pricing algorithms and dynamic pricing.

OECD. (2017b). Tackling horizontal shareholding: An update and extension to the sherman act and EU competition law—Note by Einer Elhauge. [Online] December, 2017, from https://one.oecd.org/document/DAF/COMP/WD(2017)95/en/pdf.

Prof. Dr. Kirchdörfer, R. (2011). Definition of the family business. [Online] January, 2011, from https://www.familienunternehmen.de/media/public/pdf/definition-familienunternehmen/fus_lexikon-familienunternehmen_prof-rainer-kirchdoerfer.pdf

PwC. (2017). Asset & Wealth Management Revolution [Online]. https://www.pwc.com/gx/en/asset-management/asset-management-insights/assets/awm-revolution-full-report-final.pdf.

Robyn, C. & Danah, B. (2018). Isomorphism through algorithms: Institutional dependencies in the case of Facebook [Online]. February, 2018, doi:10.1177/2053951718757253.

Shimizu, K. (2017). Technological development in automotive industry and transformation in corporate governance system. *Journal of Governance & Regulation, 6*(3), 46–54.

Chapter 4
Heterogeneity of Institutional Investors, Longevity, and Corporate Governance: The Case of Merck KGaA and Merck & Co.

Abstract Long-established corporations are generally considered stable. By examining the cases of the German pharmaceutical company Merck KGaA and US subsidiary Merck & Co., Inc. (MSD), this study examines the factors that influence corporate longevity. These two companies were once the same, with each originating from the Merck family drug store in Darmstadt, Germany, 350 years ago in 1668. Merck established its American branch in 1885. However, the two companies were separated during the First World War in 1917 because MSD was expropriated at this time. Our hypothesis is that Merck achieved a more stable longevity than MSD on the basis of share price. However, the standard deviation of the share price of Merck ($28.27) is higher than that of MSD ($15.10) over a 20-year period, thus indicating that the share price of Merck significantly fluctuated; this contradicts our expectations for the longevity also stability of the long-established corporation. The main cause of this difference might be the capitalization of Merck of 30% of their capital as a limited partnership with shares in 1995. We hypothesize that a structural governance change occurred in both corporations owing to two aspects: on the one hand, emphasizing the legal status of the general partner in Merck's family ownership structure and on the other, showing the heterogeneity of institutional investors in MSD's ownership structure for MSD.

Keywords Corporate Governance · Merck KGaA · Merck & Co. (MSD) · Longevity · Heterogeneity

4.1 Introduction

In this study, we consider the factors that affect the corporate governance of German Merck Kommanditgesellschaft auf Aktien (KGaA) (limited partners with tradable shares; referred to as the "Merck") (Merck KGaA, 2019) and the US Merck & Co. (Merck Sharp and Dohme or "MSD") (MSD Homepage, 2019). The three factors

K. Shimizu, *ESG Investment, German Industrie 4.0, and Blockchain*,
https://doi.org/10.1007/978-981-95-6927-4_4

considered are: (1) the top-management structures, (2) the ownership structures, and (3) the longevity of both corporations, see Table 4.1 (Jeroen et al., 2017). The final point is an important aspect that influences the top-management (governance) and ownership structures. The concept of longevity, which was proposed by Herbamas, is based on a coordination-oriented approach as opposed to a result-oriented approach (Habarmas, 1981). The ownership structure was formed characteristically by two elements: the free-float and family ownership. Moreover, the coordination-oriented approach (longevity) can be represented as a form of family ownership. The result-oriented approach can be described as a form of the free-float. The top-management (governance) structure referred to in the annual reports depends on German and US company laws.

The conceptual diagram in Fig. 4.1 shows the three perspectives of the historical transition (indicated by an arrow in the Fig. 4.1) of Merck and MSD, as well as their ownership structure and governance. The middle of the diagram shows the expropriation of MSD during the First World War in 1917.

Therefore, a structural governance change occurred at Merck and MSD regarding their management structure (the legal entity), ownership structure, and longevity. In general, Japanese management concepts are influenced mainly by two countries: Germany and the United States.[1] It is not only in the figurative sense to say that the "bones of Japanese management are made from Germany, and the meat is from America." Occasionally, we also attempt to understand the cross-cultural influences that create organic development for Japanese management concepts, organized pre- and post-the Second World War.

Table 4.1 Structure of this paper

<table>
<tr><th colspan="5">Structure of this paper; Corporate Governance of Merck and MSD</th></tr>
<tr><td rowspan="2">(1) Management structure</td><td colspan="4">German legal entity; KGaA (in German language): Limited partnership with share</td></tr>
<tr><td colspan="4">American legal entity; corporation, incorporation.</td></tr>
<tr><td rowspan="2">(2) Ownership structure</td><td>Free-float</td><td rowspan="2">(3) Longevity</td><td>Heterogeneity of institutional investors</td><td>Result-oriented approach</td></tr>
<tr><td>Family ownership</td><td colspan="2">Coordination-oriental approach</td></tr>
</table>

Created by the author

[1]Before the Meiji era (around the 1850s), Japan was highly influenced by Germany after the Japanese translated 'Kaitai Shinsho (New Text on Anatomy), 'which was written in German during the Edo period. After 'Sakoku' (national isolation policy), Japan was opened to American Black Ships and started trading with America.

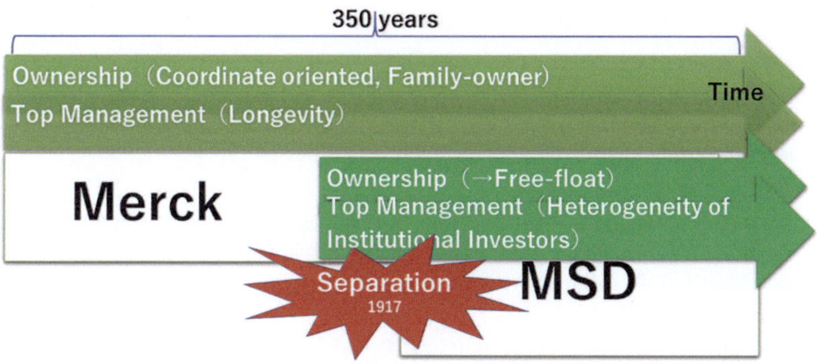

Fig. 4.1 Conceptual diagram of this paper. (Created by the author)

4.2 The Legal Entities of Merck and MSD

4.2.1 Merck (Merck KGaA)

Merck KGaA, a German multinational chemical and pharmaceutical manufacturer, was established in 1668. The present head of the Merck family is from the 13th generation, and the company has approximately 52,000 employees in around 70 countries (Merck KGaA 2019). The founder, Friedrich Jakob Merck, acquired Angel Pharmacy in Darmstadt, Germany, in 1668 and was given the privilege of establishing and running the pharmacy, which is the world's oldest operating chemical and pharmaceutical company and one of the largest pharmaceutical companies in the world.

"Merck" was originally an E. Merck open business company (Offene Handelsgesellschaft [OHG]). This business entity is usually used for family-owned companies and start-up companies. Merck changed its business entity status to a publicly traded partnership in 1995 [Masaru, 2015], and this change marked the company's most important transition. Considering the development of this form of business entity, a publicly traded partnership (Kommanditgesellschaft [KG]) is assumed from a partnership company. Table 4.2 summarizes the three most common legal entity types in Germany: sole proprietorship (individuals), partnership, and corporation. Merck KGaA is a special mix of a general partnership and a "stock company." Therefore, their control and governance structures are also separated between these two legal entities. The general partner of Merck, namely, Frank Stangenberg-Haverkamp, is part of the 13th generation. In 1984, he was elected to E. Merck KG's Board of Partners and has been the Chairman of the Board ever since (Carsten et al., 2018).

Table 4.2 List of legal entity types in Germany

Germany
Individuals
Einzelunternehmen: Individual entrepreneur, sole proprietorship; only professional services, agriculture and forestry as well as small commercial businesses
Partnerships (Personengesellschaften)
Offene Handelsgesellschaft (OHG): Translation in English "open business company," *general partnership*: No minimum capital, unlimited liability of partners
Kommanditgesellschaft (KG) limited partnership + general partnership
Corporation
Gesellschaft mit beschränkter Haftung (GmbH): Company with limited liability, limited liability company (LLC) (US); minimum equity €25,000.
***Kommanditgesellschaft auf Aktien (KGaA)**: Similar to limited partnership with shares (frank Wooldridge); Merck has been transformed their entity in 1995. Although it is a company limited by shares, the KGaA has at least one general partner whose liability is not limited. Merck listed in 1995.
Aktiengesellschaft (AG): Literally "stock company," corporation (US); minimum seven minimum capital €50,000

Source: Courtland and John (2015) and part of Wikipedia U.S.A. (2019) (Created by the author)

4.2.2 MSD (Merck & Co.)

MSD was established by E. Merck OHG from Germany in 1890 (Merck KGaA, 2019). George Merck, who was just 24 years old and was from Darmstadt, Germany, became a general partner for the American regional office in 1887 (Williams, 1983). As reparation during the First World War, Merck lost its overseas operations, and the expropriation of Merck & Co (see in Table 4.3) was eventually undertaken in accordance with the Trading with the Enemy Act of 1919 (Merck KGaA, 2019). The headquarters was located in New York City. Today, the company operates in more than 140 countries, and the number of its employees has grown to approximately 86,000. MSD shares were included in the Dow Jones Industrial Average in 1979. This is one of the main research questions why MSD is growing differently to Merck.

In this paper, we use "MSD" for the US company and "Merck" for the German company. The abbreviation MSD stands for the two main founders. The "S" stands for Mr. Sharp, and "D" stands for Mr. Dohme, who emigrated from Germany when he was 15 years old.

4.3 Financial Comparison of the Two Companies

Table 4.4 compares the financial statements of Merck and MSD (2017, currency: mil $). In 2017, MSD received approximately $40 billion in revenues compared with approximately $20 billion for Merck, whereas the net incomes for MSD and

Table 4.3 List of legal entity types in the United States of America

America
Individuals
Sole proprietorship: Owner has unlimited personal liability for the business's financial obligations.
Partnerships
General partnership: All partners have unlimited liability, meaning their personal assets are at risk to mistakes made by other partners. Profits and losses directly to the partners and are taxed at individual rates.
LP, limited partnership: Limited partners have limited liability (liable only for the amount of their investment), and general partners have unlimited liability. = > German KGaA is similar to this entity with shares, which is traded in the financial market.
Corporation
Corporation, incorporated: Unlimited number of shareholders, no limits on stock classes or voting rights, ownership and management are separated. However, the ownership structure influences a corporation's strategies. The share block holder, such as institutional investors, affects corporation's decisions.

Source: Courtland and John (2015) and part of Wikipedia U.S.A. (2019) (Created by the author)

Table 4.4 Financial statements of Merck and MSD for 2017

	Merck KGaA	MSD
Revenue (mil $)	17,166	40,122
Net income (mil $)	2912	2394
Revenue per share	39	33
	Balance sheet (mil)	
Total assets	39,896	87,872
Total liabilities	24,211	53,536
Shareholders' equity	15,683	34,336
	Cash flows (mil)	
	3020	6447
Market cap (bil $)	41.32	203.93

Source: Mergent Online, Yahoo Finance U.S. (Created by the author)

Merck are similar at $2.3 billion and $2.9 billion, respectively. This result is due to the decrease in the net income of MSD owing to income tax obligations compared with Merck. The net income of MSD is usually more than double the 5-year average. Furthermore, the balance sheet shows that MSD is almost double the size of Merck.

The market cap of MSD is approximately five times more than Merck. According to K., Gugler (2001), macroeconomic evidence is given in the share-market capitalization to Gross Domestic Product (GDP) ratio across 14 OECD countries in 1994 (average of 50.2%) (Klaus, 2001). The highest figures are in the United Kingdom (107%) and Switzerland (106%). Japan recorded 86%, the United States recorded 61%, and Germany recorded 24%. However, the country with the biggest GDP in 2017 is the United States at $19,362 billion, followed by Japan in third place at

$4,884 billion and Germany in fourth place at $3,651 billion. Share market capitalization is an important factor for understanding the influence of corporate governance in each country.

4.4 Ownership Structures of Merck KGaA and MSD & Co

4.4.1 Merck (Merck KGaA)

Given the structural differences between an AG (corporation) and a KGaA (an LP with shares), there are also major differences between the two legal forms in terms of liability and management.

In 1995, Merck KGaA (Darmstadt) was publicly listed. The strategy for their growth process and the increasing capital requirements reflected the authorized capital (100% of capital) of general partners at Merck KGaA, which became a corporation (29.7%of issued capital) with general partners (70.3%). With a volume of DM 2.4 billion, this represented the largest public offering in German history at that time. The general partner (family), namely, E. Merck KG, Darmstadt, Germany, holds the capital majority (70.3%). Figure 4.2 describes the governance structure of Merck.

E. Merck KG holds 70.3% of the total capital of Merck KGaA, and the remainder is held by other shareholders (29.7%) as share capital. E. Merck KG is excluded from the management of business activities. This is an important aspect of the *"Separation of ownership and control."*

Figure 4.2 shows the governance structure of Merck KGaA. In simple terms, Merck KGaA has two different legal entities. E. Merck KG holds 70.3% of shares, and Merck AG (Aktiengesellschaft, see Table 4.3; "stock company" or corporation) holds the remaining 29.7% of shares for trading in the capital market (share price of EUR 88.28 in June 2019). The three primary committees are the executive board (all execution), supervisory board (executive board at Merck corporation), and the board of partners and its committees (executive board at E. Merck KG, which is the general partners, and the limited liability of partners).

4.4.1.1 Executive Board Merck KGaA

Four members who are general partners with no equity interest (executive board) manage the business activities. Stefan Oschmann is the chairman of the board. These members are selected by E. Merck KG. The executive board is jointly responsible for the entire management of the company with E. Merck KG. They inform the board of partners and the supervisory board about the progress of the business and the situation of the company at least quarterly.

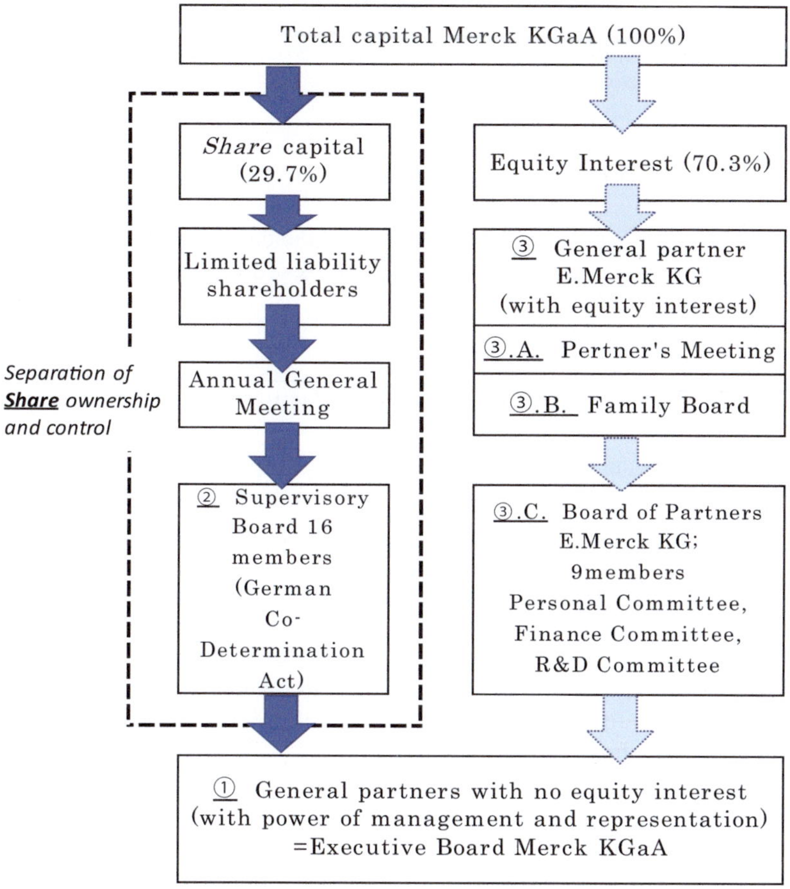

Fig. 4.2 Governance structure of Merck KGaA (a Limited Partnership with shares). *Source*: Merck KGaA, 2009, 2011 Annual Report, Corporate Governance (Created by the author)

4.4.1.2 Merck AG (29.7%)[2]

Wolfgang Büchele has served as the chairman of the supervisory board since 2016. The supervisory board is usually responsible for monitoring other departments, e.g. the executive board. However, compared with a normal German stock corporation, the role of the supervisory board of a limited partnership with shares (Kommanditgesellschaft auf Aktie; KGaA) is limited because the members of the executive board is personally liable partners and are responsible for the entire

[2] Since capitalization in October 1995, they have increased their capital several times. The small capital increase occurred between 2003 and 2008 and it was the stockoption program for directors. In 2014, they increased their Limited Partnership with shares in condition to 2:1. It formed the current ownership structure at 70.3% of the total capital of E. Merck KG and the other shareholders (29.7%) by Merck KGaA

management; this is similar to a Japanese executive board. The supervisory board is not responsible for appointing and dismissing general partners or for regulating the terms and conditions of their contracts; this is the responsibility of E. Merck KG.

4.4.1.3 E. Merck KG (70.3%)

E. Merck KG is the most powerful institution of the company. It holds four institutions, namely, partner's meeting, family board, board of partners, and the executive board.

4.4.1.4 The Partners' Meeting Gathers All Partners of E. Merck KG

Such as for the annual general meeting of a share corporation, and elects members of the family board.

4.4.1.5 The Family Board

Is responsible for the strategic direction of all Merck companies and shows the entrepreneurship capability of the Merck family as owners. Its members are elected by the partners' meeting and are also members of the Merck family (Die Stiftung für Familienunternehmen in Deutschland 2019).

4.4.1.6 The Executive Board of E. Merck KG

Frank Stangenberg-Haverkamp is the chairman of *the executive board of E. Merck KG*, and the board of partners has nine members. The board of partners is responsible for the executive board in its management of the company. It informs itself about the business matters of Merck KGaA. The executive board requires the approval of E. Merck KG for transactions that are considered the group's nonordinary business activities. The E. Merck KG Board of Partners normally meets four times a year. The members of the Merck KGaA Executive Board are invited to all meetings of the board of partners.

The board of partners has three committees in place: the personnel committee, the finance committee, and the research and development committee (Ouchi, 1981).

- *Personnel Committee*. This committee has four members and meets at least twice a year. The personnel committee is responsible for decisions concerning the members and former members of the executive board. The chairman of the committee regularly informs the board of partners of its activities.
- *Finance Committee*. This committee has four members and holds at least four meetings a year. The finance committee is responsible for analyzing and discuss-

ing the annual financial statements, the consolidated financial statements, the reports of the auditor, and the half-year financial report.

- *Research and Development Committee*. This committee has four members and holds at least two meetings a year. It is responsible for reviewing and discussing the research activities of the healthcare and life science/performance materials business sectors.

4.4.2 MSD & Co

E. Merck opened its New York branch in 1887. However, the official establishment of the first office in the United States was in 1891. Merck KGaA and MSD were separated during the First World War in 1917, and MSD was expropriated. In the same year, George Merck traveled from Germany to New York City to oversee company operations. A federal judge who was recently appointed the alien property custodian announced that the United States would seize Merck & Co., and Merck stock was sold at a public auction in the same year (MSD Homepage, 2019).

The ownership of MSD by institutional investors is now approximately 78% (NASDAQ, 2019). In contrast to German Merck KGaA, which is 70.3% family-held, MSD is widely held. The top four shareholders are several institutional investors: (1) Vanguard Group (7.65%), (2) BlackRock Fund Advisers (4.67%), (3) SSgA Funds Management (4.34%), and (4) Capital Research (3.36%) (all together sum up with four major institutional investors to 20.02%). Also, it's 29.4% in the top ten institutional investors (MarketScreener, 2019).

The approach of MSD to corporate governance is a part of corporate responsibility. They categorized their governance explanation in the category of ethics and transparency on the MSD website. The corporate governance of MSD is based on the so-called "one-tier" model[3] (see below Fig. 4.3) and comprises a board of directors with 12 external directors and a chief executive officer (Kenneth C. Frazier), four committees and an executive committee (11 members).

In *Modern Corporation and Private Property* by Berle and Means, the ownership structure of American corporations is described as widely held (Berle, A. and Means, G., 1932) and gave us the impression of "separation of ownership and management." However, in recent years, several researchers have proposed questions regarding this empirical image.

However, in recent years, several researchers have proposed questions regarding this empirical image. Rafael La Porta suggests that large corporations in 27 wealthy economies are typically controlled by families or the state (Rafael La et al., 1999). This question is also considered in this paper. Merck KGaA and MSD have different

[3] 'One-tier' means that all directors (both executive and nonexecutive directors) form one board, particularly in the United States. 'Two-tier' means that executive directors simply perform the duties they are assigned, and the job of supervision is undertaken by a separate board, namely, the supervisory board (all nonexecutive directors), similar to that in Germany.

Fig. 4.3 Structure of corporate governance for MSD. *Source*: (MSD Homepage, 2019), (Created by the author)

histories in the development of their ownership and governance structures. Indeed, since the ownership of MSD formed with the heterogeneity of institutional investors, this also shows that those shareholders can exert influence over a company even with a small holding (e.g. the largest institutional investor Vanguard Group with 7.65% in comparison with Merck KGaA with 70.3%).

4.5 Conclusion

4.5.1 Families Or the State?

On the basis of our calculations from share price data for 20 years, the standard deviation of the share price of Merck ($28.27) is higher than that of MSD ($15.10). We thought that the main factor for this difference could be the 30% capitalization of Merck and the change of its legal entity to limited partnership with shares (KGaA) in 1995. MSD was publicly listed in 1917 when it was expropriated. Federal Judge A. Mitchell Palmer[4] had to sell Merck shares at a public auction under the US Government's Alien Property Act. MSD was needed to supply the US Army and Navy with millions of vaccines throughout the First World War (1914–1918). We expect that MSD was listed at this time, according to Palmer's public offering as a

[4]He is best known for the Palmer Raids during the Red Scare of 1919–1920.

Table 4.5 Balance of listed company ownership

Country	Individual	Institutional investors	Banks and government	Holding company	Foreign
Germany	17	15	17	39	12
Japan	*20*	*21*	*23*	*28*	8
America	51	41	3	0	5

Source: Bob Tricker, "Corporate governance: principles, policies, and practices," p. 148. (Created by the author)

punishment in 1917. Therefore, MSD has approximately 100 years of experience, is more widely capitalized than Merck KGaA, which has only 25 years of experience.

Table 4.5 shows the ownership in listed companies in Germany, Japan, and the United States. In the United States, individuals and institutional investors account for 92% of shareholders; in Germany and Japan, the proportions are only 32% and 41%, respectively. Tricker (2015) suggests that the United States is an exception compared with other countries. In Germany and Japan, many listed companies are held within corporate groups. The shares are held by many dispersed external shareholders; therefore, the directors need to respond to their varied expectations (Bob, 2015, p. 148).

Rafael La Porta suggests that large corporations are typically controlled by families or the state (Rafael La et al., 1999). We also investigate German share ownership by using Monopolekommission and find that family ownership may offer a possible solution and a countervailing power against the majority ownership of the state (Monopolkommission, 2016; Shimizu, 2019).

Very interesting moment could be going in "The Unseen Revolution" by Peter Drucker 2017, which suggests it is something no one notices. Furthermore, Berle and Means consider that the separation of ownership and control occurs only in the United States and not worldwide. However, "the unseen revolution" could be something happening by the emergence of the heterogeneity of institutional investors driven by workers' pension funds.

Recently, the American Academy of Management has been discussing the type of influence that national capitalism will bring to society. For example, government agencies such as The California Public Employees' Retirement System and Japan's Pension Fund Investment Management are hybrid organizations of their respective governments. Ownership has been greatly affected by hybrid organizations that are gaining power owing to the excess in global money supply from central banks (the ECB (European Central Bank) and the FED (Federal Reserve Bank)). Furthermore, investments using the stock index, such as exchange-traded funds (ETFs), have no room for human access owing to the use of high-speed computer calculations (dehumanization); however, the stock price is still increasing.

4.6 Review Questions

1. What was the original relationship between Merck KGaA and Merck & Co. (MSD)?
2. When and why did the two companies become separated?
3. What does the concept of "corporate longevity" mean in this study?
4. What is the legal form "KGaA" (Kommanditgesellschaft auf Aktien) that Merck KGaA adopted in 1995?
5. How does a KGaA differ from a normal joint-stock company (AG)?
6. What role does E. Merck KG (the Merck family company) play in the structure of Merck KGaA?
7. How does the supervisory board of Merck KGaA differ from that of a typical corporation?
8. How can we understand the relationship between family ownership and stock market investment in Merck KGaA's governance structure?
9. What types of investors are the main shareholders of MSD?
10. How did the separation of ownership and management become an established system in corporate governance?
11. How has this system influenced the relationship between shareholders and managers?
12. Compare the theories of Berle and Means (1932) and Rafael La Porta (1999) on corporate ownership. How do they explain the international diversity of corporate governance?
13. How might the difference in stock price volatility (standard deviation) between Merck and MSD reflect their different ownership structures (family vs. institutional)?
14. Compare the financial size of both companies (sales, assets, and market capitalization). What main characteristics can be observed?
15. Which type of ownership—family-owned or institutionally owned—is more suitable for maintaining long-term corporate value, and why?

References

Berle, A., & Means, G. (1932). *The modern corporation and private property*. Retrieved from https://archive.org/download/in.ernet.dli.2015.216028/2015.216028.The-Modern.pdf

Tricker, B. (2015). *Corporate governance: Principles, policies, and practices*. Oxford University Press.

Carsten, B., Michael, K., Hermann, S., & Joachim, S. (2018). *Merck: Von der Apotheke zum Weltkonzern* (Auflage: 2 ed.). C.H. Beck.

Courtland, B. L., & John, T. V. (2015). *Business in action*. Pearson.

Die Stiftung für Familienunternehmen in Deutschland und Europa. (2019). *DIE UNTERNEHMER-MEINUNG*. Retrieved from https://www.familienunternehmen.de/de/die-unternehmer-meinung/dr-frank-stangenberg-haverkamp-und-johannes-baillou

Drucker, P. (2017). The Pension Fund Revolution, 25 September 2017.

Habermas, J. (1981). Theorie des kommunikativen Handelns.

Jeroen, N., Miguel, M., & Sophie, M. (2017). Governance implications of attracting external equity investors in private family firms. *Academy of Management Perspectives*, 1–55.

Klaus, G. (2001). *Corporate governance and economic performance*. Oxford University Press.

MarketScreener. (2019). *Shareholders by Merck & Co.* Retrieved from https://www.marketscreener.com/MERCK-AND-COMPANY-13611/company/

Masaru, Y. (2015). *Deutschlands groesste Familienunternehmen*. NTT Publishment.

Merck Geschaeftbericht 2011. (2011). *Merck KGaA*. Retrieved from http://berichte.merckgroup.com/2011/gb/corporategovernance/corporategovernancebericht/merckkgaa.html?cat=n

Merck KGaA. (2019, June 28). *Historical milestones.* Retrieved from Merck celebrates 350th anniversary: https://www.merckgroup.com/content/dam/web/corporate/non-images/press-releases/2018/feb/en/Milestones-of-our-history-EN.pdf

Monopolkommission. (2016). *Wettbewerb 2016.* Abgerufen am 2018.

MSD Homepage. (2019). *History of Merck & Co.* Retrieved from https://www.merck.com/about/our-history/home.html

Nasdaq. (2019, May 8). *Merck & Company, Inc. ownership summary*. Retrieved from https://www.nasdaq.com/symbol/mrk/ownership-summary

Ouchi, W. G. (1981). *Theory Z, how American business can meet the Japanese challenge*. Addison-Wesley Publishing Company Inc.

Rafael La, P., Florencio, L.-d.-S., & Andrei, S. (1999). Corporate ownership around the world. *The Journal of Finance*. Retrieved from https://www.jstor.org/stable/pdf/2697717.pdf?refreqid=excelsior%3A010568a59393d847356679501f3de6f7

Shimizu, K. (2019). *"Pricing Game" for tacit collusion and passive investment.* CEUR Workshop. Retrieved from http://ceur-ws.org/Vol-2348/paper23.pdf

Wikipedia U.S.A. (2019). *List of legal entity types by country*. Retrieved from https://en.wikipedia.org/wiki/List_of_legal_entity_types_by_country#Japan

Williams, H. (1983). *American chemical industrie, a history*. Garland Publishing, Inc.

Part II
ESG Investment and German Industrie 4.0

Chapter 5
The Digital Transformation of Work and ESG as a New Framework for Social Security

Abstract Digital transformation (DX) has reshaped work environments and significantly influenced national social security systems. This chapter argues that digitalization and Internet-based restructuring have generated a pluralization of value systems, expanding the investment concept of centrally managed social security funds toward a broader, network-oriented ESG framework. The Environmental (E) dimension is defined by the global shift toward energy transition and carbon neutrality, while the Social (S) dimension now encompasses employees, customers, and communities, forming a complex social network. Governance (G) has evolved from a traditional corporate mechanism into a dynamic vector linking E and S and guiding societal structures.

Comparing Germany, Japan, and the United States, the chapter investigates how economic growth outcomes are distributed through indicators such as unemployment rates and Gini coefficients. The findings highlight a "productivity paradox," showing that digitalization increases productivity but does not necessarily improve social welfare. This underscores the importance of evaluating economies through multidimensional value indicators rather than monetary metrics alone.

By analyzing the balance between social security benefits and contributions, the study reveals the expansion of ESG investment concepts and the gradual shift from institutional public fund management toward individual self-investment. Challenges such as the limited flexibility of national pension funds and risks of greenwashing are also identified.

As a response, the chapter demonstrates the effectiveness of blockchain technology—exemplified by the European Battery Passport—in ensuring transparency and data integrity. These findings suggest that ESG investment frameworks possess strong resilience and adaptability, even in periods of economic turbulence.

Keywords Digital transformation (DX) · Platform monopoly · Social security · SRI (Socially Responsible Investment) · ESG (Environment, Social, Governance)

K. Shimizu, *ESG Investment, German Industrie 4.0, and Blockchain*,
https://doi.org/10.1007/978-981-95-6927-4_5

5.1 Introduction

In particular, Japan, Germany, and the United States (with occasinal reference to China and Russia) are undergoing a reorganization of their employment structures and welfare state models under distinct institutional designs and strategies for introducing IoT technologies, notably those associated with Industrie 4.0.

As discussed in the previous chapter, ESG investment provides a framework that incorporates the three elements of environment, social, and governance into investment decision-making. It is grounded in the relationship of trust and fiduciary duty between investors and trustees. In the management of pension funds and similar institutions, it is essential to balance returns and risk avoidance while also fulfilling social responsibility. From this perspective, social security systems can also be understood as mechanisms through which governments conduct ethical investments on behalf of individuals.

Two basic approaches underlie this framework:

- Negative screening (SRI), which excludes companies deemed socially inappropriate, and
- Positive screening (ESG), which favors companies that excel in environmental and social contributions (Shimizu 2018).

Through the use of IoT technologies and the European Data Space initiative, the transparency of such investments has been improving. However, concerns remain about "greenwashing"—the superficial presentation of environmental responsibility in corporate reporting and social security disclosures. To ensure the credibility of ESG information, mechanisms that strengthen resistance to data manipulation are necessary, and third-party audits and blockchain-based technologies have been attracting growing attention as potential solutions.

In Japan, in response to an aging society, efforts are being made to build a human-centered IoT society, known as Society 5.0, with a particular focus on elderly care, healthcare, and energy policy. As exemplified by Toyota's "Woven City" smart city project, the digitalization of social welfare systems and the use of technology to support daily life have become central pillars of a sustainable society (JIL 2006).

In Germany, under the Industrie 4.0 national strategy, the advancement of production systems through AI and robotics and the reskilling of workers are progressing rapidly. These initiatives, led by major corporations such as Siemens and SAP, are increasingly spreading to small- and medium-sized enterprises (Mittelstand). Industrial policy (Industrie 4.0) and labor policy (Arbeit 4.0) are being closely coordinated, fostering a social mechanism that enables adaptation to technological transformation on a broad scale.

In contrast, the United States is promoting technology adoption through market-driven logic, characterized by rapid automation and a widening income gap. Although its social security system has not fully kept pace with these changes, technology is beginning to play a complementary role in certain aspects of social protection within an efficiency-oriented approach.

In summary, the trajectories of these three countries are deeply rooted in their respective historical contexts and institutional frameworks.

- Japan emphasizes life-support-oriented IoT solutions as a response to aging and resource constraints, integrating them with social welfare systems.
- Germany seeks to integrate technology and institutional reform, testing the limits of social adaptability.
- The United States, while pursuing efficiency through market mechanisms, continues to face challenges in terms of social inclusion (Fig. 5.1).

ESG investment is an international framework based on the United Nations' Principles for Responsible Investment (PRI) and is closely linked to the long-term growth strategies of both corporations and nations. It goes beyond traditional Corporate Social Responsibility (CSR), aiming instead to realize a form of ethical capitalism.

This global trend is also connected to the idea of Effective Altruism (EA), which can be seen as a decentralized form of social security. However, incidents such as the collapse of FTX have led to the misconception and undervaluation of its underlying philosophy. The key challenge ahead lies in designing a sustainable coexistence between rapidly evolving technology and human society. Achieving greater equality and social inclusion will require institutional reforms that keep pace with technological innovation, combined with comprehensive social welfare and education policies. As John Kenneth Galbraith once argued, community-based strategic giving organizations grounded in effective altruism could emerge as a new counterbalancing force against market-driven self-interest. Digital Transformation (DX) has fundamentally altered the concept of work and the nature of labor, reconstructing the surrounding economic ecosystem. With the progress of DX, the

Fig. 5.1 Conceptual framework of this study. (Created by the author)

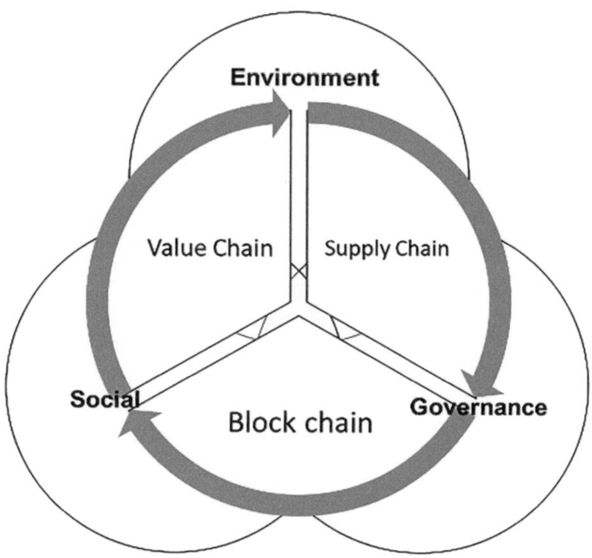

conventional metrics used to evaluate capital and labor—such as GDP—are also changing. The rise of digital platform corporations such as Google, Apple, Facebook, Amazon, and Microsoft (GAFAM) has given them economic influence surpassing that of many states, thereby affecting the very foundations of democracy and market economies (Lessig, 2006). As a result, productivity gains have become increasingly difficult to measure using traditional indicators, while inequality—reflected in the Gini coefficient—has continued to rise over the past three decades. Nevertheless, in advanced economies, especially in the Nordic countries, mechanisms such as basic income guarantees have helped mitigate inequality and promote a stronger emphasis on work–life balance.

5.2 Evaluating the Quality of Economic Growth

To assess whether the benefits of economic growth are being widely shared across society, it is necessary to examine both the unemployment rate and the Gini coefficient. Figure 5.2 presents the unemployment rate, while Fig. 5.3 illustrates the Gini coefficient.

5.2.1 Analysis of Changes in the Unemployment Rate

In 2000, the average unemployment rate among the five countries was relatively high at approximately 6.1%. However, as labor market conditions improved in each country, the rate declined steadily, reaching around 3.4% by 2010 (Fig. 5.2).

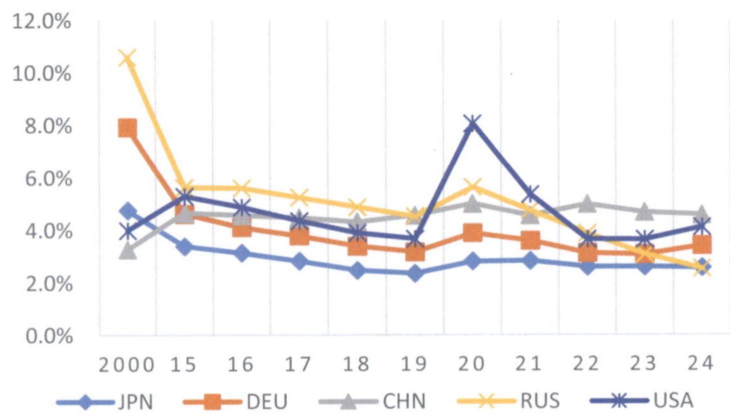

Fig. 5.2 Trend of unemployment rate of Germany (DEU), Japan (JPN), China (CHN), Russia (RUS), United States (USA). (Created by the author)

Overall, unemployment rates have shown a consistent downward trend over the past two decades. Although some countries experienced a temporary increase in unemployment around 2020 due to the COVID-19 pandemic, the situation has since stabilized. In particular, Russia and Japan have maintained long-term declines in unemployment, reflecting stable employment environments.

Based on regression analysis using unemployment data from 2014 to 2024, the following country-specific trends were identified (*T*-value: significance of the slope/*P*-value: statistical significance):

- Japan (JPN): Significant downward trend (slope = −0.057, $T = -2.42$, $P = 0.038$) → Annual decrease of approximately 0.06 percentage points, indicating a stable labor market.
- Germany (DEU): Significant decline (slope = −0.101, $T = -3.01$, $P = 0.015$) → Likely reflecting effective labor policies and a robust industrial structure.
- Russia (RUS): Most pronounced decrease (slope = −0.327, $T = -6.21$, $P = 0.00016$) → Rapid labor market changes possibly influenced by external factors such as sanctions and demographic shifts.
- China (CHN): Slight increase (slope = +0.014) but statistically insignificant ($T = 0.72$, $P = 0.488$) → Overall stability without a significant trend.
- United States (United States): Large fluctuations due to the COVID-19 shock, showing no clear long-term trend (slope = −0.090, $T = -0.71$, $P = 0.496$) → Highly sensitive to business cycle variations.

This analysis suggests that changes in unemployment rates are strongly influenced by institutional and external factors specific to each country. A large *T*-value and $P < 0.05$ indicate statistically significant changes.

Overall, the average unemployment rate has generally declined, reflecting ongoing labor market improvement across these economies (World Bank, 2020a).

5.2.2 Analysis of Changes in the Gini Coefficient

National redistribution through taxation and other fiscal mechanisms generally helps to alleviate inequality, as measured by the Gini coefficient. Figure 5.3 presents the Gini indices (World Bank estimates) for Germany, Japan, China, the Russian Federation, and the United States. [World Bank, 2020b].

Regression analysis of time-series data on Gini coefficients for each country revealed the following trends:

- Japan: The Gini coefficient increased by approximately +0.36 points per year (slope = 0.363), with $T = 2.64$, $P = 0.058$, and $R^2 = 0.64$, indicating a moderately significant upward trend in inequality.
- Germany: An annual increase of +0.16 points (slope = 0.162) was observed, showing strong statistical significance ($P = 0.00018$) and a high explanatory

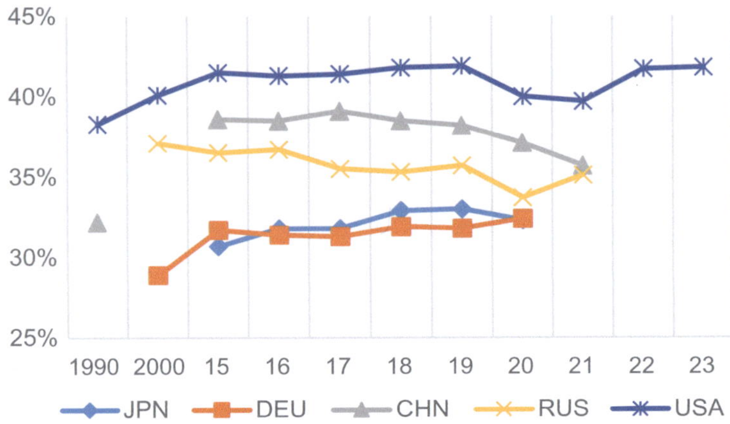

Fig. 5.3 Trend of Gini coefficient of Germany (DEU), Japan (JPN), China (CHN), Russia (RUS), United States (USA). (Created by the author)

power (R^2 = 0.95). This may reflect changes in redistribution mechanisms and increased labor-market flexibility.
- China: The Gini coefficient rose significantly by +0.18 points per year (slope = 0.180, $T = 3.15$, $P = 0.0198$, $R^2 = 0.62$), statistically confirming a widening income gap.
- Russia: A slight decline of −0.11 points per year (slope = −0.113) was found, though the significance level was marginal ($T = -2.39$, $P = 0.054$, $R^2 = 0.49$). This may be influenced by inflation in the 1990s, income redistribution policies, and changes in statistical reporting after the 2000s.
- United States: A gradual increase of +0.08 points per year (slope = 0.082) was statistically significant ($T = 3.09$, $P = 0.0129$, $R^2 = 0.51$), confirming a clear trend of widening inequality.

Overall, most countries show an upward trend in Gini coefficients, with inequality growth in Germany, China, and the United States. In contrast, countries like Russia experienced a temporary decline in inequality, but the *P*-value near 0.05 warrants careful interpretation.

Figures 5.2 (Unemployment Rate) and 5.3 (Gini Coefficient) reveal contrasting trends: since 2000, unemployment has generally decreased, while income inequality has increased. Paul (2017) similarly observed the global progression of inequality, describing it as a paradox in which "employment rises while inequality widens." Paul argues that capitalism is a system that "creates prosperity" but does not necessarily "create happiness." The root cause, he suggests, lies in the disconnection between human rationality (System II) and emotion (System I), as well as between individual interests and the public good (Paul, 2017).

Annual Change represents the slope coefficient (β) from a linear regression of each indicator on time (year), using complete-case analysis (see Table 5.1).

Table 5.1 Time-series analysis of unemployment and Gini coefficient (Created by the author)

Country	Unemployment (Annual change, β)	Gini coefficient (Annual change, β)	Summary
JP Japan	**−0.095** (significant)	**+0.036** (marginally significant)	Unemployment decreased; inequality slightly increased
DE Germany	**−0.202** (significant)	**+0.16** (marginally significant)	Unemployment decreased; inequality increased
CN China	**+0.061** (significant)	**+0.18** (significant)	Unemployment increased; inequality increased
RU Russia	**−0.309** (significant)	**−0.11** (borderline significant)	Unemployment decreased; inequality slightly decreased
US USA	**+0.012** (no significant change)	**+0.08** (significant)	Unemployment stable; inequality increased

Economist Erik Brynjolfsson et al. (2018) identifies this phenomenon as the "paradox of growth" observed since the 1990s. Despite rapid advances in AI technology, aggregate productivity has not improved as much as expected—a phenomenon known as the "AI Productivity Paradox." There are four main explanations:

1. Expectations of outcomes have outpaced actual implementation;
2. The benefits of AI are difficult to capture in macroeconomic statistics such as GDP;
3. Profits are concentrated among a limited number of firms and individuals; and
4. Institutional and organizational adaptation has lagged behind technological change.

As with earlier technological revolutions, the positive effects of AI take time to materialize. Human resource development and organizational reform are indispensable for realizing its full potential (Brynjolfsson et al., 2018).

Frey and Osborne (2013) estimated that about 47% of all US jobs are at risk of automation, while Rotmanarchive (2013) highlighted the "decoupling" between rising productivity and stagnant employment, contributing to widening inequality (Frey & Osborne, 2013; Rotmanarchive, 2013).

Furthermore, Jurgens (2024) notes that diversification and automation in the automotive industry have increased the number of relatively simple, low-wage jobs. In Germany and Japan, institutional constraints make employment adjustment difficult, whereas in the United States, labor market flexibility (Lerning curve) is much higher. This indicates that inequality is shaped not only by technological content but also by how societies and organizations adopt and manage technology (Jurgens, 2024).

As technologies based on statistical inference—such as AI—become more pervasive, mutual understanding and consensus-building among social stakeholders are becoming increasingly vital. In this process, the role of social media and other digital platforms in shaping public consensus cannot be overlooked.

Ulrich (2025) also points out that in the implementation of Industrie 4.0, the flexibility of labor in final-assembly processes has improved. Yet, the expansion of simple assembly work may be contributing to rising income inequality, even as unemployment remains low (Ulrich, 2025).

With the rapid evolution of AI and big data, computers now play an increasingly central role in economic systems. In both corporate management and public policy, computer-based simulations are commonly used to model economic systems and forecast future impacts. However, such IT-driven value creation is difficult to measure using traditional productivity metrics such as "one price per good" or "output equals value."

Digital technologies increasingly create value not through tangible goods, but through information and experiential value.

Meanwhile, in physics, Richard Feynman (1982) raised the question of whether nature itself can be precisely simulated by computers. He argued that classical computers are incapable of reproducing quantum phenomena—such as superposition and nonlocality—accurately (Feynman, 1982).

To overcome this limitation, the quantum computer was proposed: a fundamentally new type of computation that uses the very laws of quantum mechanics to perform calculations, enabling a truer simulation of nature.

This paradigm shift also offers new insights for economics. The eighteenth-century economist Adam Smith famously proposed that through the "invisible hand," individual self-interest could lead to social benefit—for example, a baker rising at 4 a.m. to bake bread acts out of self-interest, yet ultimately contributes to the well-being of others.

However, in the modern digital economy, informed by AI and quantum simulation, we are beginning to see complex interactions that transcend the binary of "selfish" versus "altruistic." The pursuit of profit, when integrated with information technology and natural law, may give rise to a higher form of social harmony (Paul, 2017).

5.3 The History of Work and Technology

This section provides a brief overview of the historical background of technological development relevant to this study.

5.3.1 Manual Labor, Mechanization, and Electrification

- 1700—Manual Labor:

  When production was dominated by physical labor, flexibility was at its highest. Flexibility is one of the main driving forces behind digitalization. Therefore, in the final stage of digital transformation, the essence of manual production—its flexibility—is reincorporated, enabling multi-product, small-lot production.

- 1800—The First Industrial Revolution (Mechanization):

The transition from manual labor to machine looms powered by water and steam marked the First Industrial Revolution. The steam engine, fueled by coal, made coal mining a highly profitable industry. The introduction of new technologies brought profound social changes, including the emergence of a stronger middle class in Britain.

- 1900—The Second Industrial Revolution (Electrification and Mass Production):

The Second Industrial Revolution introduced electric-powered mass production based on division of labor. Electrification of factories and the establishment of modern production lines led to significant economic growth and productivity gains worldwide.

Modern assembly lines were first adopted on a large scale in Chicago's meat-packing industry. However, mechanization also displaced many factory workers, increasing unemployment.

In response, the Factory Acts were enacted by the British Parliament in the early nineteenth century to regulate industrial working conditions, focusing particularly on child labor hours and moral welfare.

In the automotive industry, remarkable changes occurred: at Ford's Highland Park Plant, the labor hours required to produce one Model T dropped dramatically—from 1776 h in 1908 to just 53 h in 1926—representing a 33-fold productivity improvement due to automation through single-purpose machines and assembly lines.

However, since the 1970s, productivity growth has slowed. In the body assembly process, labor hours decreased only modestly—from 33 h in 1926 to 29 h in 1974, an improvement of just 14%.

5.3.2 The Evolution of Digitalization (Data-Driven Production)

- 1960—The Third Industrial Revolution (Digital Revolution):

Also known as the Digital Revolution, this period emerged after the two World Wars. During the Cold War, rapid computation was essential for cryptographic analysis and missile trajectory calculations.

One of the most significant innovations of this era was the Internet, originally developed by the US Department of Defense as "ARPANET," a communication network designed to maintain contact in the event of nuclear war.

The concept of Artificial Intelligence (AI) also originated at the 1956 Dartmouth Conference, where researchers explored methods of constructing artificial creativity [Yulia, 2020].

- 1970—Conceptual Development of AI:

  The concept of AI consists of three essential components: intelligent machines that can learn, process, and ultimately create natural language.

  The learning process can be illustrated using an Euler diagram divided into four interrelated segments. Deep learning forms the smallest circle at the core, embedded within machine learning, which in turn is a subset of AI.

  The Euler diagram highlights the interdependence among these domains and clarifies the boundaries that structure the accumulation of digital resources.

- 1980—Machine Learning:

  Machine learning emerged as a subfield of AI focused on teaching computers how to learn without explicit programming.

  It requires three main components: datasets, features, and algorithms. Features are data attributes that indicate what the machine should pay attention to.

  While human decision-making is boundedly rational, the use of better-curated datasets and more appropriate feature selection can make such systems more efficient than humans in specific tasks.

- 2010—Deep Learning:

  Deep learning represents an advanced form of machine learning, inspired by the neural networks of the human brain.

  Since around 2015, central processing units (CPUs) have evolved to incorporate the capabilities of graphics processing units (GPUs). GPUs increase computational productivity by performing parallel processing and offloading heavy tasks from the CPU.

  Deep learning employs multi-layered neural network architectures that convert abstract, non-linear datasets from one layer to another. GPUs enable such operations to be executed faster, cheaper, and more powerfully, giving "eyes" to intelligent machines.

  This breakthrough may be compared to the "Cambrian Explosion" in evolution, when organisms developed vision about 500 million years ago. Technologies such as autonomous driving have thus become a tangible reality.

A new generation of technologies is now shaping the future of AI—most notably, multimodal AI, which integrates multiple forms of data such as images, audio, and text to make advanced judgments, mimicking the human ability to perceive the world through the five senses.

The term "multimodal" derives from the Latin multus ("many") and modalis ("mode"). In the context of human cognition, multimodality refers to the ability to encode and decode external environments using multiple sensory modalities. As humans continue to scientifically analyze their own cognitive functions, the capabilities of AI are expected to expand even further.

5.4 Previous Studies

This section briefly reviews existing research related to digitalization and work.

5.4.1 M. Krzywdzinski and C. Gerber

Krzywdzinski and Gerber observe that the platform economy may exacerbate social inequality. To test this hypothesis, they employed several methodological approaches and conducted an empirical analysis based on case studies of 15 cloud work platforms in the United States and Germany. Their findings indicate that the weaker the social safety net, the more likely platform work provides flexible income sources under highly precarious conditions. Conversely, the stronger the social safety net, the greater the workers' market power over the platform (Krzywdzinsk & Gerber, 2020).

There exists a significant relationship between social safety nets and pension systems. In Anglo-Saxon countries, pensions are typically invested in financial markets to ensure adequate benefits for retirees. In contrast, on the European continent, pensions and social security benefits are provided mainly through government agencies, making taxation indispensable for system sustainability.

5.4.2 Eurofound

This study focuses on the expansion of non-standard employment throughout the 2010s. It found that between the mid-1990s and 2007, both temporary contracts and self-employment grew relatively strongly. The most relevant form of non-standard work is associated with digitalization—namely, the so-called "gig economy," where work is mediated through online platforms traded on stock markets (Azar et al. 2017). In fact, the stock market anticipates strong growth potential in this sector. However, it is premature to assume significant employment growth as a direct consequence of current digital platforms (Ezrachi & Stucke 2015). A potential barrier to such future expansion is the replacement of labor by new technologies.

Social protection constitutes a crucial issue for digital platform work. The gig economy often categorizes income and other conditions as temporary contract-based work. Consequently, questions of labor law and social protection depend largely on whether workers are classified as self-employed or as employees. According to the European Social Insurance Platform, comprehensive reforms of social protection are not yet deemed necessary for digital platforms that facilitate virtual work (Eurofound, 2017).

The potential for future growth in the gig economy remains highly speculative. Although digital platform work has grown rapidly in recent years, many well-known platforms (such as Airbnb) derive their profits from capital gains on stock markets. However, several factors could constrain future expansion. Since part of these profits arises from tax arbitrage, tax authorities are increasingly aware of and responsive to such issues, particularly as the scale of tax-related challenges continues to grow. Airbnb's business model benefits from several tax advantages, such as reduced rates on business property taxes and value-added taxes (Houlder, 2017).

5.5 Trends in Social Security Benefits and Contributions in Japan, Germany, and the United States

5.5.1 *Interpretation of Regression Analysis: Social Security Benefits (2005–2023, 19 Years) and Contributions (2004–2022, 19 Years)*

Based on standardized OECD coefficients, this regression analysis reveals striking differences among Germany, Japan, and the United States regarding the design and direction of their social security systems (see Table 5.2).

Japan exhibits a consistent upward trend in both benefits and contributions. In particular, the regression coefficient for contributions is +0.185, with an exceptionally high determination coefficient (R^2 = 0.984), indicating a strong temporal correlation and institutional stability. This reflects demographic aging and efforts toward fiscal sustainability. While benefits also show an upward trend, the lower R^2 value (0.366) suggests that annual fluctuations are influenced by economic conditions and policy adjustments.

Germany's contributions show a positive slope of +0.154 with a moderate explanatory power (R^2 = 0.711), reflecting adaptive reforms responding to demographic change and employment restructuring. In contrast, benefits exhibit a slight decline (β = −0.064) with a low R^2 (0.122), indicating high year-to-year variability and the absence of a clear long-term trend.

Table 5.2 Trends in social security benefits and contributions in Japan, Germany, and the United States (Created by the author)

Country	Category	Slope (β)	Intercept	Regression equation	R^2 (Coefficient of determination)
Germany DE	Benefits	−0.064	135.165	$\hat{y} = -0.064x + 135.165$	0.122
	Contributions	+0.154	−309.725	$\hat{y} = 0.154x - 309.725$	0.711
Japan JP	Benefits	+0.110	−203.604	$\hat{y} = 0.110x - 203.604$	0.366
	Contributions	+0.185	−371.527	$\hat{y} = 0.185x - 371.527$	0.984
United States US	Benefits	+0.129	−243.186	$\hat{y} = 0.129x - 243.186$	0.497
	Contributions	−0.049	99.243	$\hat{y} = -0.049x + 99.243$	0.074

In the **United States**, benefits and contributions move in opposite directions. While benefits show a mild upward trend ($\beta = +0.129$), contributions display a slight decline ($\beta = -0.049$) with an extremely low explanatory power ($R^2 = 0.074$). This suggests that US social security contributions are more sensitive to short-term economic cycles and policy shifts than to long-term institutional or demographic factors.

Overall, these findings highlight the distinct welfare state models of each country: Germany's "Rhineland model," Japan's "hybrid model," and the United States' "market-oriented model." Japan demonstrates consistent institutional expansion, Germany gradual adaptation through reforms, and the United States a flexible, policy-driven system responsive to economic fluctuations.

5.5.2 Stock Prices, Consumption, and Savings

Generally, stock prices and consumer behavior are highly correlated. As a leading indicator, stock prices often reflect overall economic expectations. For example, in the United States—where households hold a high proportion of their assets in equities—changes in stock market performance significantly influence consumption behavior. Similarly, the EU and Japan show comparable behavioral tendencies in response to major stock index fluctuations.

While stock prices influence consumption, how do they relate to the opposite concept—savings? Why do households save money? Although motivations vary, one major reason is long-term financial security, particularly through pension savings for post-retirement life. According to the OECD, household saving rates in G7 countries (Canada, France, Germany, Italy, Japan, the United Kingdom, and the United States) have steadily declined for over two decades. In Germany, for instance, the presence of comprehensive social insurance systems—covering healthcare and substantial retirement pensions—fulfills most post-retirement needs. Consequently, Germany's saving rate remains relatively stable compared with other G7 nations (OECD, 2020; Poterba, 1993).

5.6 Sustainable Investment

According to estimates by Deutsche Bank, the total value of global financial assets in 2014 was approximately USD 242 trillion. This capital market—worth USD 242 trillion—was distributed across various financial instruments such as stocks, bonds, and derivatives. Based on calculations by the Japan Securities Research Institute (2006), the combined total value of equities, bonds, and derivatives amounted to around USD 112.8 trillion. Using updated statistics from the World Federation of Exchanges (WFE, 2017), the estimated market value of securities in 2017 was approximately USD 169.5 trillion, comprising equities (USD 85.3 trillion), bonds (USD 76.2 trillion), and derivatives (USD 8 trillion).

Globally, the total value of sustainable investment assets reached USD 30.7 trillion at the beginning of 2018. This amount represented roughly 19% of all securities assets at that time (Deutsche Bank Research, 2014; Japan Securities Research Institute, 2006; World Federation of Exchanges, 2026).

According to the Global Sustainable Investment Review (2018, 2022), Europe continued to lead in sustainable investment, accounting for 46% of global assets, though its share has stagnated since 2016 (previously around 53%). The United States maintained a nearly steady share at 39%, while Japan recorded remarkable growth—tripling its sustainable investment assets—following the Government Pension Investment Fund (GPIF)'s adoption of ESG principles into its investment strategy (The Global Sustainable Investment Alliance, 2018).

Major sustainable investment strategies include:

- Negative/Exclusionary Screening
- Positive/Best-in-Class Screening
- ESG Integration

5.6.1 *Negative/Exclusionary Screening*

Negative or exclusionary screening refers to excluding specific sectors or companies from a fund or portfolio based on predetermined ESG criteria. As of 2016, the largest sustainable investment strategy worldwide was negative/exclusionary screening (USD 19.8 trillion), followed by ESG integration (USD 17.5 trillion). Negative screening remains the dominant strategy in Europe, where assets under this method have increased by 31% over the past 2 years, reaching USD 19.7 trillion.

In the United States, ESG integration continues to be mainstream, whereas in Japan, corporate engagement and shareholder action are the prevailing strategies (The Global Sustainable Investment Alliance, 2025).

5.6.2 *Fiduciary Duty*

According to Milton Friedman, "No one spends somebody else's money as carefully as he spends his own." Following classical economic principles, Friedman argued that corporate executives have a duty to maximize shareholders' economic returns and should not pursue other objectives that could conflict with this responsibility.

He stated that managers must operate efficiently within legal and ethical boundaries to serve shareholder interests, but charitable acts or social responsibility initiatives should not undermine profitability. Moreover, he warned that executives who pursue social objectives at the expense of shareholders "undermine the foundations of a free society" (Ethics Briefing 2013)

However, Friedman did not anticipate the rise of advanced ESG-oriented funds—such as those managed by BlackRock—that follow a "comply or explain" principle, integrating social responsibility into fiduciary frameworks (BlackRock 2017) (Madison 2020).

5.6.3 Value, Supply Chain, and Fragmented Value

In his seminal 1985 book Competitive Advantage, Michael Porter introduced the concept of the value chain, which encompasses key activities such as logistics, marketing, and human resources that together create competitive advantage. Porter suggested that firms could develop such an advantage through interrelations among these activities.

Later, Poter and Heppelmann (2014) discussed how information technology (IT) transforms product value chains. With the acceleration of digital transformation (DX), IT-driven innovations have led to significant gains in productivity and economic growth. Increasingly, IT has become an integral part of products themselves—through embedded sensors, processors, software, and connectivity—linked to "product clouds" where product data are stored and analyzed. These are known as smart, connected products [Micheal & James, 2014].

The value chain creates high product value through the interrelation of internal and external activities. As workflows become more fragmented, unique work processes are increasingly integrated via computer networks. In modern organizations, workers are not only tied to physical structures but also participate in cybernetic systems where data-driven processes and machine networks coexist (Robert, 1992).

Such digital transformation of work challenges traditional social security systems, which often cannot adapt to new forms of labor. ESG-oriented approaches therefore provide a framework for rethinking and reconstructing these institutional systems.

Porter's later concept of "shared value" further bridges the link between social impact and profitability, suggesting that ESG strategies should not be seen as opposing profit but rather as synergistic. For instance, in the power generation industry, government regulations historically determined electricity prices, granting regional monopolies that ensured steady profits. Porter's Five Forces Analysis at that time would have correctly predicted stable profitability, and many investors purchased utility stocks. However, as deregulation advanced, new competition emerged, and innovations such as solar and wind energy reshaped the market landscape (Porter et al., 2019).

5.7 Stock Market Performance During the Financial Crisis and the COVID-19 Pandemic

The COVID-19 crisis triggered a surge of investment in high-tech industries—perceived as both low-carbon and resilient. Among the world's top ten publicly listed companies by market capitalization, seven were connected to Internet and communication technologies (Microsoft, Apple, Amazon, Alphabet, Alibaba, Facebook, and Tencent).

To examine the impact of ethical standards on market performance, this study compares: the performance of 137 companies excluded by the Norwegian Council on Ethics, which oversees the Government Pension Fund of Norway, and the Global Dow Index, which represents 150 top companies selected by Dow Jones (Norges Bank, 2020; Wikipedia, 2020).

- Norwegian Council on Ethics–Exclusion List. The Government Pension Fund of Norway is a sovereign wealth fund owned by the Norwegian government. Established in 1990 as the Petroleum Fund, it manages over USD 1 trillion in assets, making it the world's largest SWF (sovereign wealth fund). It is also the largest pension fund in Europe and larger than one of the biggest US institutional investors, CalPERS. As a very large SWF, it is expected to invest in accordance with fair-trade and ethical guidelines [The Council on Ethics, 2020].
- The Global Dow. An equity index composed of 150 leading companies worldwide, selected by Dow Jones. [S&P Dow Jones Indices, 2020].

5.7.1 *The Lehman Brothers (Capital Liquidity) Crisis*

On September 15, 2008, Lehman Brothers—then the fourth-largest US investment bank—filed for bankruptcy, marking the largest corporate failure in history (assets: USD 639 billion; liabilities: USD 619 billion). The crisis stemmed from a collapse in capital liquidity, leading to simultaneous declines in stock prices across both ethical and non-ethical companies.

An analysis of 306 data points (November 3, 2008–December 31, 2010) using Excel's CORREL function revealed a correlation coefficient of 0.9524, indicating an extremely strong relationship between the excluded companies and the overall market index.

- Total return of excluded companies: +34.45%
- Total return of Global Dow: +29.80%

Hence, during liquidity crises like Lehman's, ethical screening did not significantly alter performance since market-wide shocks affected all firms almost uniformly.

5.7.2 The COVID-19 Crisis (Disruption of Logistics and Labor)

The COVID-19 pandemic, first identified in Wuhan, China, in December 2019, was declared a Public Health Emergency of International Concern by the World Health Organization (WHO) on January 30, 2020.

Unlike the Lehman crisis—which was rooted in capital liquidity—the COVID-19 shock was a real-economy crisis caused by the disruption of logistics and human mobility. This shift emphasized corporate responsibility and social sustainability, elevating the role of ESG investment derived from Socially Responsible Investment (SRI) (Knoll 2002).

The Norwegian Council on Ethics excludes companies that:

- Violate fundamental humanitarian principles,
- Engage in nuclear weapons, tobacco, or arms manufacturing, or,
- Conduct business contrary to societal ethics.

A correlation analysis using 218 data points (November 1, 2019–August 20, 2020) showed a correlation coefficient of 0.9226, lower than during the Lehman crisis.

- Total return of excluded companies: −10.23%
- Total return of Global Dow: −0.35%

These results—consistent with findings by Amélie et al. (2016)—suggest a paradigm shift in ethical investment, where ESG and SRI principles became increasingly influential during real-economy disruptions (Amélie et al., 2016).

5.7.3 Summary of Both Crises (See Table 5.3)

The Lehman crisis represented a collapse of capital liquidity, where ethical and non-ethical investments behaved similarly. In contrast, the COVID-19 crisis—a real-economy shock—revealed greater resilience among ethically and socially responsible companies.

This comparison suggests that ESG investment demonstrates greater effectiveness during real-economy crises than in pure financial liquidity crises.

5.8 Conclusion

This chapter highlights the following points:

Since 2000, while unemployment rates have fallen in many countries, income inequality (measured by the Gini coefficient) has risen—indicating that

Table 5.3 Capital liquidity and disruption of logistics and labor crisis (Created by the author)

Item	Lehman Crisis (Collapse of Capital Liquidity)	COVID-19 Crisis (Disruption of Logistics and Labor)
Period	November 3, 2008—December 31, 2010 ($n = 306$)	November 1, 2019—August 20, 2020 ($n = 218$)
Nature of the crisis	Collapse of capital market liquidity (financial system crisis)	Stagnation of the real economy (pandemic-induced halt in logistics and labor mobility)
Primary cause	Credit contraction and breakdown of financing channels	Global supply chain disruption and restrictions on human mobility
Correlation (excluded stocks vs. global Dow)	0.9524 (very high correlation)	0.9226 (moderately lower correlation)
Total return—Norwegian council on ethics exclusion list	34.45%	−10.23%
Total return—Global Dow (standard index)	29.80%	−0.35%

employment growth has not translated into equitable distribution. One reason is the "productivity paradox"—where digital transformation (DX) enhances efficiency but is not fully captured by traditional productivity metrics. Hence, economies require multidimensional value indicators beyond monetary measures, incorporating social and environmental dimensions.

In asset management, institutional investors managing public pensions increasingly combine SRI (negative screening) and ESG (positive screening) strategies under fiduciary responsibility (MSCI 2026). This hybrid approach strengthens the complementarity between public investment and individual self-investment.

Meanwhile, to prevent "greenwashing" (superficial environmental claims), efforts are underway to integrate blockchain technology into third-party audits, enhancing data integrity and traceability. The EU Battery Passport exemplifies this development.

Institutional designs also differ across countries:

- Japan links social welfare with Society 5.0 and IoT-based life support systems.
- Germany integrates Industrie 4.0 with Arbeit 4.0, uniting technology policy and labor reform.
- The United States, though efficient and market-driven, faces persistent challenges in social inclusion.

In comparing crisis responses, the Lehman crisis saw little performance difference between ethical and non-ethical investments due to liquidity collapse. However, during COVID-19, firms adhering to ESG principles demonstrated stronger resilience—the capacity to sustain performance amid disruptions in logistics and labor.

From these findings, three shared challenges emerge:

1. Institutionalizing education and reskilling programs;
2. Developing data reliability frameworks that integrate auditing with blockchain technologies;
3. Enhancing ESG integration in pension fund management.

Together, these steps can help resolve the paradox of "low unemployment yet rising inequality," improving the overall quality of growth.

5.9 Review Questions

I. Explain why the unemployment rate has fallen while the Gini coefficient has risen, using the perspectives of digital transformation (DX) and job polarization.
II. Why did the effectiveness of ESG investment differ between the Lehman crisis and the COVID-19 crisis?
III. Explain why blockchain is an effective tool against greenwashing, particularly in relation to the audit process.

References

Amélie, C., Darné, O., & Fouilloux, J. (2016). The impact of screening strategies on the performance of ESG indices. *HAL*. Retrieved from https://hal.archives-ouvertes.fr/hal-01344699

Azar, J., Schmalz, M., & Tecu, I. (2017). Anti-competitive effects of common ownership. *FTC microeconomics conference 2015*. Retrieved November, 2018, from https://www.ftc.gov/system/files/documents/public_events/633601/azar_presentation.pdf

World Bank (2020a). *Free and open access to labour statistics*. Retrieved from https://ilostat.ilo.org/data/

World Bank (2020b). *GINI index*. Obtenido de Data: https://data.worldbank.org/indicator/SI.POV.GINI

BlackRock. (2017). *PUBLIC POLICY; common ownership*. Retrieved from https://www.blackrock.com/corporate/literature/whitepaper/viewpoint-index-investing-and-common-ownership-theories-eng-march.pdf

Brynjolfsson, E., Daniel, R., & Chad, S. (2018). *Unpacking the AI-productivity paradox*. MIT Sloan. Retrieved from https://sloanreview.mit.edu/article/unpacking-the-ai-productivity-paradox/?utm_source=twitter&utm_medium=social&utm_campaign=sm-direct

Ethics Briefing. (2013). *Ethical indices*. Institute of Business Ethics. Retrieved from https://www.truevaluemetrics.org/DBpdfs/Metrics/IBE/IBE-Ethical-Indices.pdf

Deutsche Bank Research. (2014). *Mapping the world's financial markets 2014*. Deutsche Bank. Retrieved from https://etf.dws.com/DEU/DEU/Download/Research-Global/47e36b78-d254-4b16-a82f-d5c5f1b1e09a/Mapping-the-World-s-Financial-Markets.pdf

Eurofound. (2017). *Non-standard forms of employment: Recent trends and future prospects*. Eurofound. Retrieved from https://www.eurofound.europa.eu/publications/customised-report/2017/non-standard-forms-of-employment-recent-trends-and-future-prospects

Ezrachi, A., & Stucke, M. E. (2015). *Artificial intelligence & collusion: When computers inhibit competition*. Retrieved November, 2018, from https://illinoislawreview.org/wp-content/uploads/2017/10/Ezrachi-Stucke.pdf

Feynman, R. P. (1982). Simulating physics with computers. *International Journal of Theoretical Physics, 21*(6/7) Retrieved from https://s2.smu.edu/~mitch/class/5395/papers/feynman-quantum-1981.pdf

Frey, C. B., & Osborne, M. (2013). The future of employment: How susceptible are jobs to computerisation? . Retrieved from https://www.oxfordmartin.ox.ac.uk/publications/the-future-of-employment

Houlder, V. (2017, January 2). Airbnb's edge on room prices depends on tax advantages. *Financial Times*, 47. Retrieved from https://www.ft.com/content/73102c20-c60e-11e6-9043-7e34c07b46ef

Japan Securities Research Institute. (2006). 世界の資本市場と証券・資産運用ビジネス. Japan Securities Research Institute. Retrieved from http://www.jsri.or.jp/publish/topics/pdf/0702_01.pdf

The Japan Institute for Labour Policy and Training; JIL. (2006). Social security costs borne by companies. Retrieved from: https://www.jil.go.jp/foreign/labor_system/2006_9/world_01.html

Jurgens, U. (2024). *Automation and its impact on productivity and workerslessons from the history of the car industry*. Weizenbaumournal of the Digital Society. Retrieved from https://ojs.weizenbaum-institut.de/index.php/wjds/article/view/5_2_4/193

Knoll, M. S. (2002). *Ethical screening in modern financial markets: The conflicting claims underlying socially responsible investment*. Business Lawyer, 57 Bus. Law. 681. Retrieved August 29, 2017, from https://www.researchgate.net/publication/279647341_Ethical_Screening_in_Modern_Financial_Markets_The_Conflicting_Claims_Underlying_Socially_Responsible_Investment

Krzywdzinsk, M., & Gerber, C. (2020). *Varieties of platform work. Platforms and social inequality in Germany and the United States*. Weizenbaum Series #7. Retrieved May, 2020, from https://www.weizenbaum-institut.de/media/Publikationen/Weizenbaum_Series/Weizenbaum_Series_7_Krzywdzinski_Gerber.pdf

Lessig, L. (2006). *Code: And other laws of cyberspace, version 2.0*. Basic Books.

Madison, C. (2020). Externalities and the common owner. Retrieved from https://papers.ssrn.com/sol3/papers.cfm?abstract_id=3378783

Micheal, P., & James, H. E. (November, 2014). "How smart, connected products are transforming competition". Harvard Business Review, pp. 64–88. Obtenido de https://hbr.org/2014/11/how-smart-connected-products-are-transforming-competition

MSCI. (2026). *MSCI ESG criterion*. Obtenido de https://www.msci.com

Norges Bank. (2020). The list of companies that are excluded from the Government Pensions Fund Global. Norway. Retrieved from https://www.nbim.no/en/the-fund/responsible-investment/exclusion-of-companies/

OECD. (2020). *OECD Data*. Retrieved from Saving rate: https://data.oecd.org/socialexp/social-spending.htm#indicator-chart.

Paul, G. D. (2017). The limits of the market: The pendulum between government and market. *Oxford Academic*. https://doi.org/10.1093/acprof:oso/9780198784289.003.0002

Porter, M. E., Serafeim, G., & Kramer, M. (2019, October 16). Where ESG Fails. (I. Investor, Ed.) Retrieved from https://www.institutionalinvestor.com/article/b1hm5ghqtxj9s7/Where-ESG-Fails

Poter, M., & Heppelmann, J. (November, 2014). "How smart, connected products are transforming competition". Harvard Business Review, pp. 64–88. Obtenido de https://hbr.org/2014/11/how-smart-connected-products-are-transforming-competition

Poterba, J. M. (1993). *International comparisons of household saving*. University of Chicago Press. Retrieved from http://www.nber.org/chapters/c8869

Robert, A. E. (1992). Michael Porter's competitive advantage and business history. *Business and Economic History, 21*(Second Series), 231–233. Retrieved from https://www.jstor.org/stable/23703225?seq=1

Rotmanarchive, D. (2013, June 12). *How technology is destroying jobs.* Retrieved July 2020, from https://www.technologyreview.com/2013/06/12/178008/how-technology-is-destroying-jobs/

S&P Dow Jones Indices. (April, 2020). *The global dow methodology*. Obtenido de file:///C:/Users/shimi/Downloads/methodology-dj-the-global-dow.pdf.

Shimizu, K. (2018). The dilemma between "comply or explain" and SRI, ESG methodology; transitional terminology. : Organisational Information Ethics. Obtenido de https://www.researchgate.net/publication/322594259_The_Dilemma_between_comply_or_explain_and_SRI_ESG_methodology_transitional_terminology

The Council on Ethics. (2020). *sovereign wealth fund (SWF)*. Obtenido de https://etikkradet.no/en/

The Global Sustainable Investment Alliance. (2018). *The global sustainable investment review 2018.* Retrieved from http://www.gsi-alliance.org/wp-content/uploads/2019/06/GSIR_Review2018F.pdf

The Global Sustainable Investment Alliance. (2025). *The global sustainable investment review 2022.* Retrieved from https://www.gsi-alliance.org/members-resources/gsir2022/

Ulrich, J. (2025). Automation and its impact on productivity and workers; lessons from the history of the car industry. *The Weizenbaum Journal of the Digital Society*. https://doi.org/10.34669/wi.wjds/5.2.4

Wikipedia. (2020). *The global dow*. Retrieved from https://en.wikipedia.org/wiki/The_Global_Dow

World Federation of Exchanges. (2026). WFE annual statistics guide (volume 4). The United States. Retrieved from https://www.world-exchanges.org/our-work/articles/wfe-annual-statistics-guide-volume-4

Yulia, G. (April, 2020). *Artificial intelligence vs. machine learning vs. deep learning: Essentials.* Obtenido de https://serokell.io/blog/ai-ml-dl-difference

Part III
Industrie 4.0 as Networking Technology

Chapter 6
Technological Development in the Automotive Industry and Transformation in the Corporate Governance System

Abstract This chapter examines how the development of a ubiquitous society—where everything is connected to Internet—and the spread of IoT are transforming industrial structures, governance systems, and value creation. Focusing on the automotive industries of Germany, Japan, and the United States, it analyzes how advancements in Industrie 4.0, IoT, embedded production technologies, and AI are restructuring corporate governance and accelerating cross-industry integration. The analysis is based on six perspectives presented in the chapter: 1. GDP Trends for Manufacturing Industries, 2. Industrie 4.0 & Governance Structure, 3. Viewpoint Relating to IoT Usability, 4. Smiling Curve & Value Chain Network, 5. Production System & Embedded System, 6. Cyber-Physical System & ICT-Related Development.

These perspectives include: (1) changes in manufacturing's share of GDP, (2) governance structures shaped by Industrie 4.0, (3) IoT usability, (4) value allocation illustrated by the smiling curve, (5) production systems and embedded technologies, and (6) the development of CPS—bridging virtual and physical domains—and ICT. While manufacturing has declined in the United States, Japan and Germany have maintained stability through policy measures. The smiling curve serves as a framework for explaining differences in value distribution across virtual and real spaces.

From an institutional viewpoint, the chapter compares how Japan's Keiretsu, Germany's Konzern, and the Anglo-American capital-efficiency-oriented model have shaped supply chain management. It further argues that the networking of IoT technologies and embedded systems lies at the center of industrial transformation. Car navigation systems function as exemplary CPS, supporting autonomous driving technologies and serving as intelligent nodes in connected mobility networks.

These structural changes necessitate new governance frameworks. Drawing on Norbert Wiener's cybernetics, the chapter concludes that designing a "cybernetic map" is essential for navigating rising complexity. Ultimately, institutional safeguards that support diversity, innovation, and fair competition—and that protect

K. Shimizu, *ESG Investment, German Industrie 4.0, and Blockchain*,
https://doi.org/10.1007/978-981-95-6927-4_6

"freedom of design"—are indispensable for building adaptive and resilient industrial ecosystems.

Keywords Embedded system · Internet of Things (IoT) · Supply chain management (SCM) · Automotive navigation system · Cybernetics

6.1 Introduction

In this chapter, we examine the deployment of IoT and Industrie 4.0 technologies in the automotive industry, referencing the discussions of Ray Kurzweil (Kurzweil, 2010) and J.K. Galbraith (Galbraith, 1967) as the theoretical framework (Table 6.1).

Ray Kurzweil, a futurist born in February 1948, proposed the concept of the "technological singularity," a turning point in which the development of artificial intelligence will fundamentally surpass human intelligence.

He states: "Today, the computer in your mobile phone is 1 million times cheaper, 1 million times smaller, and 1000 times more powerful than the computers in the 1970s. This represents a billion-fold increase in computing power per dollar. Furthermore, similar technological advancements will occur again over the next 25 years." (Kurzweil, 2010/2/11).

Kurzweil's perspective suggests that by 2020, a single computer would surpass the intelligence of a human, and by 2045, artificial intelligence will have the capacity to exceed the collective intelligence of humanity by a factor of one billion. This projection fundamentally redefines the relationship between technology and humanity.

Table 6.1 Discussion matrix

Discussion Matrix	*1) GDP Trends for Manufacturing Industries*	*2) Industrie4.0 & Governance structure*	*3) Viewpoint relating to IoT usability*
Germany	Stable	Konzern	Producer-centric (smart FACTORY)
Japan	Upward	Keiretsu	Producer-centric (smart FACTORY)
USA	Downward	Anglo-American model	User-centric (smart CAR)
Discussion matrix	*4) Smiling curve & value chain network*	*5) Production system & embedded system*	*6) Cyber physical system & ICT related development*
Germany	INDUSTRIE 4.0, up & middle stream	MQB (non-linear)	Privacy-oriented digital transformation
Japan	Middle stream, hybrid motor and EV	TNGA (non-linear architecture)	AI oriented
USA	Down stream, software defined vehicle (SDV)	IoT (incremental and dynamic innovation)	EV (tesla) with GAFAM platformars

Author's given

Various industries, including the automotive sector, are experiencing a paradigm shift driven by the evolution of the Internet and ICT technologies. Traditional static organizational structures are being transformed into dynamic network structures.

Major automotive manufacturers are increasingly connecting their production equipment through the Internet, aiming to reduce inventories, optimize costs, and shorten lead times.

This shift signifies a move away from conventional hierarchical frameworks to more flexible, networked configurations, where data flows seamlessly across systems, enabling real-time decision-making and process optimization. Rapid decision-making across diverse applications increasingly relies on abundant, heterogeneous, and real-time data. The number of Internet-connected devices has grown exponentially, from approximately 6 billion in 2017 to over 14 billion in 2022 (Zhuming et al., 2025).

This conceptualization of IoT is rooted in a producer-centric perspective—especially within manufacturing activities such as supply chain management (SCM)—which emphasizes low inventory levels and lean production systems. Conversely, consumers remain dependent on suppliers, while producers respond by carefully monitoring demand through marketing research.

J.K. Galbraith (1967) discusses the role of "technostructure" that corporations hold within society. He points out that corporations, through the development of advertising, create artificial demand (dependency effect), thereby significantly enhancing their social influence. This increase in demand, in turn, serves as a catalyst for large-scale mass production of goods and services. Furthermore, Galbraith warns that the rise of corporate-driven consumption may lead to the neglect of traditional values such as fairness, beauty, trust, and loyalty (Galbraith, 1967).

Henry Mintzberg views technostructure as a bureaucratic control system aimed at stabilizing organizational operations. Moreover, he argues that technostructure not only functions internally to manage activities but also extends its influence to external networks, promoting efficiency and economic objectives across entire industries and national economies.

In the aforementioned perspective, the advancement of network technology is seen as a means to link bureaucratic de facto standards (standardized tasks) in both tangible and intangible forms. Tangible linkages include assembly line management in manufacturing, warehouse management systems, and logistics tracking systems. On the other hand, intangible linkages are exemplified by cloud-based knowledge management systems using hyperscalers. It calls for a reimagining of organizational dynamism through the "design of freedom," promoting efforts to transcend fixed value systems and move toward alternative, pluralistic value creation.

Recent advancements in IoT and Industrie 4.0 have enabled the collection of high-precision data that can be transferred to HD maps. This capability is crucial for autonomous vehicles that require centimeter-level accuracy. Traditional GPS systems can only achieve a maximum accuracy of around 100 cm, which is insufficient for the precision required by autonomous vehicles. While advanced correction technologies such as DGPS (Differential GPS) can enhance GPS accuracy, they require specialized infrastructure.

On the other hand, connected vehicles play a vital role in updating HD maps by transmitting real-time, high-precision data collected through onboard sensors. This data includes details such as road gradients, precise corner points, and obstacle locations, which conventional GPS cannot capture. Consequently, a "real-time geospatial data network" is established, significantly improving the navigation accuracy of autonomous vehicles.

- *Tangible Tasks in Map Creation*: Sensor installation and management, data collection vehicle operations.
- *Intangible Tasks in Map Creation*: User Interface (UI) design, real-time data analysis.

Currently, mobile mapping applications provide navigation for various modes of transportation, such as vehicles, pedestrians, and bicycles, all displayed on a single map. However, these apps primarily rely on GPS data, which can result in several meters of inaccuracy. For autonomous vehicles that demand centimeter-level precision, conventional GPS-based systems are inadequate. Additionally, when overlaying routes for cars, pedestrians, and bicycles on the same map, visual confusion and misinterpretation can occur if spatial requirements specific to each mode (e.g., the boundary between roadways and sidewalks, the exact position of crosswalks) are not accurately represented. Thus, the integration of high-resolution geospatial data through HD maps becomes essential. This allows for precise differentiation between vehicle, pedestrian, and bicycle data, thereby enhancing both safety and accuracy.

6.2 Reason for Structural Governance Change

6.2.1 Manufacturing Value Added and GDP Share Comparison: China, Germany, Japan, UK, and United States

According to the World Bank (2023), Fig. 6.1 below presents data on the value added by manufacturing activities and their percentage share in GDP for major countries—China, Germany, Japan, and the United States—at 5-year intervals from 2000 to 2021. As shown in Fig. 6.1, China recorded the highest share of manufacturing activity, with 32.1% in 2000, which declined to 27.5% in 2021 (a decrease of 4.6 percentage points). A general downward trend can also be observed in Japan (22.5–20.9%), the United Kingdom, and the United States (15.2–10.4%). However, Germany demonstrated a somewhat stable trend, with 20.3% in 2000 and 18.7% in 2021 (World Bank, 2025).

The differences observed among these major countries can be attributed to monetary policies driven by changes in the industrial structure of the United States.

Key automotive manufacturers have relocated their production bases to regions with lower labor costs, such as China. Japan's Toyota, Germany's Volkswagen, and

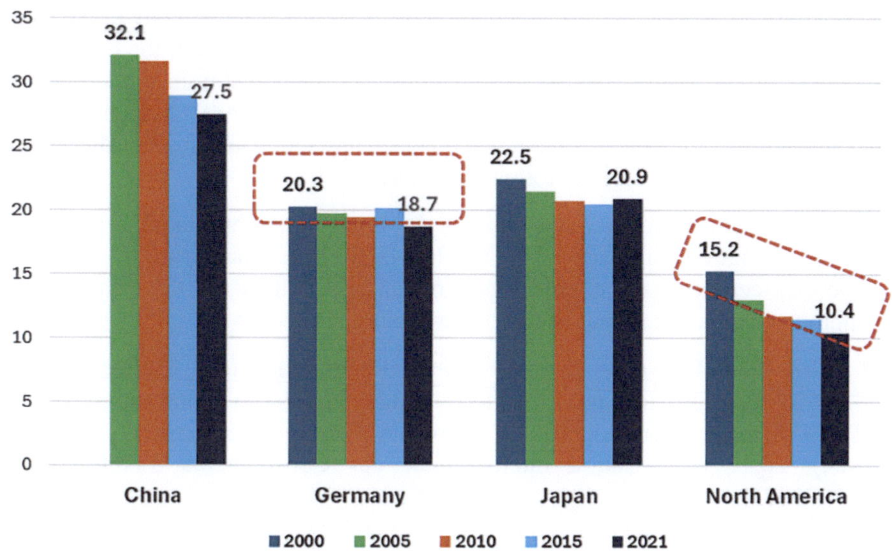

Fig. 6.1 Comparison of the share of value added by manufacturing activities in GDP among major countries (Created by the author)

the United States' Ford have shifted their manufacturing facilities abroad, thereby generating profits through foreign direct investment from their home countries (National Accounts Main Aggregates Database, 2026). As of 2025, the appreciation of the US dollar—caused by interest rate differentials between the United States and Japan (and the EU) since 2022—is gradually being corrected by the imposition of tariffs under the Trump administration. Foundational to the global optimization of such supply chain management (SCM) systems are lean production methodologies and the Industrie 4.0 initiatives, which will be discussed below (Shintaku, 2016).

6.3 Reason for the Differences Between Downward and Stable Trends

6.3.1 Smart Factory Or Smart Car? The Concept of IoT Usability

This section, titled "Smart Factory or Smart Car? A Perspective on IoT Usability," is a substantial revision of the content presented by the author at the ICN 2016 International Conference held in Lisbon (Shimizu, 2016) and the subsequent paper "Technological Development in the Automotive Industry and Transformation in Corporate Governance System" (Shimizu, 2017).

The study focuses on the implementation methods of IoT technology in the automotive sector in Japan, Germany, and the United States, exploring the conceptual differences in IoT usability across these nations. The analytical framework incorporates the "Smiling Curve Concept" introduced by Shih (Shih, 2002), which serves as a model for illustrating the upstream and downstream dominance in the value chain.

This chapter emphasizes the following contrasts:

- The producer-centric German Industrie 4.0 implementation model, represented by the "Smart Factory" approach prevalent in Germany and Japan.
- The user-centric IoT implementation model, exemplified by the "Smart Car" approach that is prominent in the United States.

In 2023 (see Fig. 6.2), the global automotive production volume reached 92,724,668 units. Of this total, the top 18 manufacturers collectively produced 70,229,000 units, accounting for 75.7% of the global market share (Association (JAMA), 2025; Organisation Internationale des Constructeurs d'Automobiles (OICA), 2025).

This dominance by the top 18 manufacturers underscores the concentrated nature of the automotive market, leaving the remaining 24.3% (approximately 22,495,668 units) to be produced by other companies. When represented as a pie chart, this distribution clearly illustrates that the top 18 manufacturers occupy approximately three-quarters of the global automotive market, while the remaining quarter is distributed among other producers. This visual representation effectively conveys the substantial market influence exerted by the top manufacturers. In Japan, major

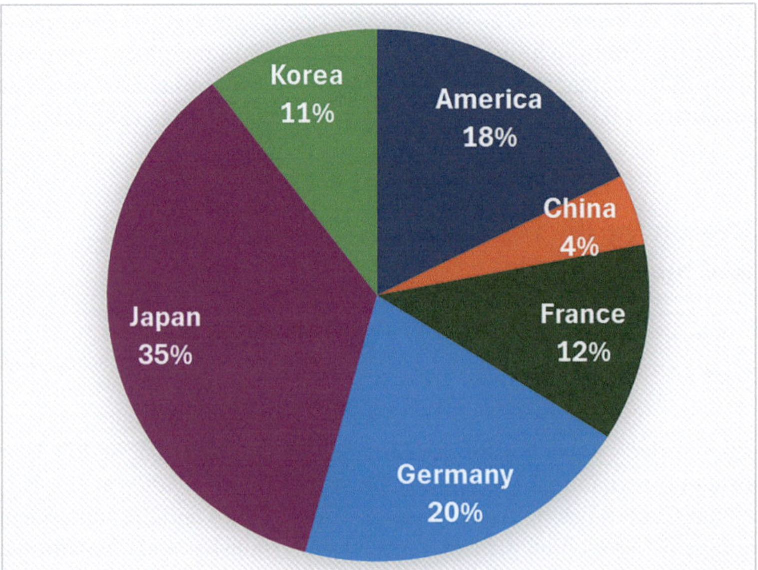

Fig. 6.2 Global automotive production volumes and market share in 2023 (Created by the author)

producers such as Toyota (over ten million units annually), Nissan, Honda, Suzuki, Subaru, Mitsubishi and Mazda are included among these top 18 manufacturers.

The global automotive market is characterized by a highly concentrated structure, dominated by Japan, Germany, and the United States. Japan leads the global automotive industry with a commanding 35% market share, significantly surpassing other countries. This dominance underscores the enduring competitiveness of Japanese manufacturing, particularly in the automotive and manufacturing sectors. This strength is further emphasized by Japan's distinctive SCM networks and production methodologies, such as the Toyota Production System, which effectively highlight its strong presence in the global market.

Germany and the United States occupy the second and third positions, with market shares of 20% and 18%, respectively. Germany's competitive edge lies in its focus on high-quality, high-value-added manufacturing and the integration of innovative production systems like Industrie 4.0, which leverage advanced machinery and digitalization. In contrast, the United States is less reliant on manufacturing and more characterized by its dominance in the ICT sector, exemplified by GAFAM, enabling significant penetration across multiple sectors as a digital powerhouse.

France and South Korea also play pivotal roles in the global automotive market. France holds a 12% share, exerting its influence through luxury goods and the automotive sector. South Korea, with an 11% share, maintains a robust presence in both electronics and the automotive industry.

Despite its vast industrial base, China accounts for only 4% of the market share in terms of manufacturer-specific production volumes at the time of these statistics. However, China's continuous investments in technological innovation and manufacturing capacity, particularly in electric vehicles and battery production, have positioned it as the world's largest EV producer. This anticipated growth presents a substantial opportunity to reshape the global automotive market structure.

6.3.2 *The Difference in GDP Trends for Manufacturing Activities Between 2015 and 2023*

Table 6.2 compares the percentage distribution of global automotive production among Germany, Japan, the United States, and other countries between 2015 and 2023. The data highlights shifts in market share over this period, indicating changes in the global automotive landscape.

Table 6.2 The difference in GDP trends for manufacturing activities between 2015 and 2023

Year/Country	*Germany*	*Japan*	*USA*	*Other*
2023	20.0%	35.0%	18.0%	27.0%
2015	19.0%	33.0%	19.0%	29.0%
Difference	**1.0%**	**2.0%**	**−1.0%**	**−2.0%**

OICA, Author's given

- **Japan**: Increased its share from 33% to 35%, maintaining its leading position in global automotive production. This growth reflects Japan's strong focus on electric and hybrid vehicle production.
- **Germany**: Expanded its share from 19% to 20%, driven by advancements in automated and high-tech manufacturing.
- **United States**: Declined slightly from 19% to 18%, suggesting a slight reduction in its production dominance amid growing competition.
- **Other Countries**: Their combined share decreased from 29% to 27%, indicating increased market concentration among the top three countries.

Overall, the data shows a trend toward consolidation in the global automotive market, with Japan, Germany, and the United States collectively accounting for 73% of total production in 2023, up from 71% in 2015. This concentration underscores the influence of major manufacturers in shaping global production patterns.

6.4 Industrie 4.0 (Smart Factories)

Germany's automotive and manufacturing industries are increasingly integrating IoT to connect production equipment and reduce inventory, costs, and lead times. These initiatives represent the early stages of smart factory development, which requires advanced Supply Chain Management (SCM) systems. According to 2022 data from CECIMO, Germany currently ranks as the world's third-largest market in manufacturing equipment sales (CECIMO, 2025). This trend is closely linked to the number of vehicles produced, as automotive parts are manufactured using industrial machinery.

Traditional automotive manufacturing has been shaped by the principles of lean production, significantly reducing excess inventory [2]. This study compares the perspectives of producers and users in the automotive sector.

Volkswagen (VW) in Germany and Toyota in Japan are both focused on optimizing production systems. However, their approaches differ:

- Volkswagen employs a module-based production system known as Modularer Querbaukasten (MQB).
- Toyota utilizes a lean-based production system, the Toyota New Global Architecture (TNGA).

The Japanese model is characterized by an industrial structure centered around Keiretsu, emphasizing long-term cooperative relationships and stability. In contrast, the German model adopts a centralized structure that prioritizes labor-management relations through Konzern-style organizational structures and modularized product designs. On the other hand, the Anglo-American model, represented by network-based companies such as GAFAM, prioritizes short-term profit maximization and capital efficiency as its core principles.

Returning to the central discussion, the combination of components in Toyota and VW's systems demonstrates differing applications of IoT technology. This variation in component integration influences the overall usability of IoT, creating divergent perspectives between producers and users.

Traditionally, systems that connect machinery have been closed systems, where hardware and control software are isolated, limiting the number of functional modules and, consequently, the system's scalability and capability. To overcome these limitations, the Internet serves as a universal infrastructure that connects all devices and systems using standard protocols such as HTTP (Hypertext Transfer Protocol) and SMTP (Simple Mail Transfer Protocol). This connectivity expands the functionality of IoT systems and specialized industrial machinery, significantly enhancing the scalability and interoperability of Industrie 4.0 systems.

Germany's automotive and manufacturing sectors are actively leveraging IoT technologies to enhance their competitive edge within the framework of INDUSTRIE 4.0. The term "4.0" represents the evolution of industrial revolutions from 1.0 to 4.0, highlighting key milestones in industrial development:

- **First Industrial Revolution (1.0)**: Initiated in the seventeenth century, it marked the introduction of mechanical production systems powered by water and steam.
- **Second Industrial Revolution (2.0)**: From the eighteenth to the nineteenth century, mass production based on division of labor and powered by electricity was introduced. The concept of large-scale assembly lines was first implemented by applying a reverse approach (engineered) to the disassembly process observed in Chicago's meatpacking plants.
- **Third Industrial Revolution (3.0)**: In the twentieth century, electronics and information technology (IT) were integrated into manufacturing, leading to further automation of production processes.

Currently, **The Fourth Industrial Revolution (Industrie 4.0)** leverages IoT and Cyber-Physical Systems (CPS) to interconnect machines, systems, and processes, enabling the optimization of manufacturing, engineering, resource utilization, supply chains, and lifecycle management. CPS, supported by low-cost sensors, bridges the physical and digital worlds. Against this backdrop, Plattform Industrie 4.0 is driving the development of the DataSpace Industrie 4.0—a global data space designed to ensure data sovereignty, security, and integrity. This initiative aims to accelerate future value creation in industrial manufacturing and foster the emergence of sustainable digital ecosystems.

In contrast, the United States has adopted a more user-oriented approach. Companies like Google have developed autonomous vehicles that integrate a variety of sensors—such as GPS receivers, laser rangefinders, video cameras, and radar—to collect big data in cyberspace. This data-driven manufacturing model emphasizes user-centric innovation, as data from physical systems is continuously fed back into the digital space for further analysis and optimization.

Globally, several countries have implemented similar national IoT strategies to enhance industrial competitiveness:

- United States: Advanced Manufacturing National Program (AMNP), (Executive Office of the President, President's Council of Advisors on Science and Technology, 2014).
- France: La nouvelle France industrielle (Siemieniuch et al., 2015).
- United Kingdom: Future of Manufacturing.
- Japanese "Society 5.0" (The government of Japan, 2017).

Germany's INDUSTRIE 4.0 is positioned as a part of this global network of high-level IoT initiatives, fostering international competition through technological innovation.

6.4.1 Segmentation of Automotive Parts and Smart Connected Products

A vehicle consists of over 30,000 components, which can be broadly categorized into five main segments: Engine, Drivetrain, Body, Chassis, and Interior/Body Interior.

- **Engine**: The heart of the vehicle, responsible for converting thermal energy into mechanical energy.
- **Drivetrain**: Transmits power from the engine to the wheels via the transmission, drive shaft, and differential.
- **Body**: Generally refers to the vehicle's hood, doors, and trunk, primarily composed of steel.
- **Chassis**: Currently defined as a structural framework that includes the suspension, steering, tires, and wheels.
- **Interior/Body Interior**: Essential components that provide comfort and functionality, including seats, dashboards, air conditioning, and audio systems.

Vehicles may also include additional features to enhance passenger comfort.

Table 6.3 The segmentation of various automotive parts. Both Volkswagen (VW) and Toyota have developed highly innovative production systems that emphasize modularity.

Table 6.3 Car parts segmentation

Car Part Segmentation	I. Engine	II. Drivetrain	III. Body	IV. Chassis	V. Interior
Toyota	*Hybrid electricity*	Drive wheels	Bonnet	Suspension	Sheets
	Hydrogen engine	Transmission	Doors	Steering	*Dash boards*
Volkswagen	*Down sized engine*	Drive shaft	Trunk	Tires	*Air conditioner*
	Ecofuel engine	Differential		Wheels	*Audio*

OICA, Author's given

The concept of modularity is widely used in studies on technical and organizational systems. A product system is considered modular when it can be decomposed into multiple components that can be recombined in various configurations.

For instance, the interior body components—such as the dashboard, air conditioning system, and audio system—can be grouped into a single modular package. This module is then connected to a centralized main computer system that is supported by widely used, cost-effective sensors, enabling enhanced system integration and connectivity.

A computer is, in fact, one of the best examples of a modular design. Common modules in computer systems include the power supply unit, processor, motherboard, graphics card, hard drive, and optical drive. These components are designed to be easily interchangeable, utilizing standardized interfaces that facilitate seamless replacement and upgrades.

M. Porter and J. Heppelmann discuss the historical transition of IT within the product value chain. They identify three waves of IT-driven transformation (Porter & Heppelmann, 2014):

First Wave (1960s–1970s): Individual activities within the value chain were automated using computers. Tasks such as order processing and invoice payments were managed through computerized systems.

Second Wave (1980s–1990s): The widespread and cost-effective adoption of the Internet-enabled ubiquitous connectivity. This phase led to significant productivity gains and economic growth, but it did not fundamentally alter the products themselves.

Third Wave (Present): IT has become an integral part of the products themselves. Sensors, processors, software, and connectivity are now embedded within products, and data generated by these products is collected and analyzed in the product cloud. This evolution has given rise to what are now known as smart connected products.

Modularization functions as a mechanism that enables fine-tuning even in products that require close integration, such as automobiles. The buffer that supports this fine-tuning allows for both product diversification and development efficiency, while the incorporation of Kaizen activities further enhances the product's competitive advantage. This evolutionary trajectory in production systems began with mass production through scientific management and assembly line methods, followed by Toyota's system of low-volume, high-variety production. Modularization, in turn, enabled a shift from these linear processes to more dynamic configurations. Ultimately, Industrie 4.0 introduced a paradigm where the direction of production processes can flexibly adapt both vertically and horizontally. This can be interpreted as a form of processual coordination akin to Japan's traditional Suri-awase (mutual adjustment) approach, where integration and flexibility coexist across highly digitized and networked manufacturing systems.

6.4.2 *Smiling Curve*

To explain the somewhat stable manufacturing trends in Japan and Germany and the declining trend in the United States as shown in Fig. 6.1, we can apply the conceptual framework of the Smiling Curve discussed in the introduction of this text. The Smiling Curve illustrates the potential for value creation across various components of the value chain, primarily within the manufacturing sector.

The Value Chain, as conceptualized by Porter, comprises two main types of activities (Porter, 1998):

- *Primary Activities*: Inbound logistics, operations, outbound logistics, marketing & sales, and services.
- *Support Activities*: Infrastructure, human resource management, technology development, and procurement.

In the Smiling Curve (Fig. 6.3), the horizontal axis represents the stages of value creation from left to right. The upper section of the curve, labeled A, C, E, corresponds to the cyberspace, where intangible assets such as patents, technology, R&D, branding, and service reputation contribute to value creation. However, these elements remain largely unrealized in physical form at this stage.

The lower section of the curve, labeled B, D, F, represents the physical space, where tangible products and components contribute to value creation. Here, the

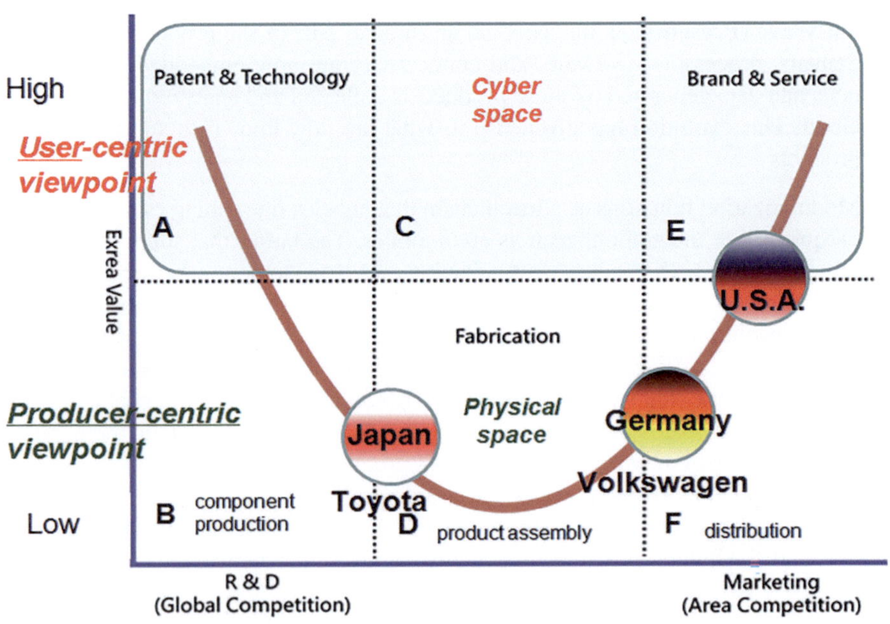

Fig. 6.3 Positions of Japan, Germany, and the United States on the smiling curve (Created by the author)

emphasis is on the materialized products, such as automotive components, where physical attributes define the economic value.

While the Smiling Curve focuses on the distribution of value along the value chain, the Value Curve provides a broader perspective on how value is created and distributed across the entire industry.

The Value Curve divides the industry into three primary stages along the horizontal axis:

- **Upstream (Component Manufacturing)**—dominated by global competition driven by technological expertise, production efficiency, and economies of scale.
- **Midstream (Product Assembly)**—involves the integration of components into finished products, where firms aim to optimize cost efficiency and production flexibility.
- **Downstream (Distribution and Sales)**—characterized by regional or localized competition, focusing on brand power, marketing channels, and logistics capabilities.

In the context of value creation, the term "value" encompasses two critical dimensions:

6.4.2.1 Monetary Value

This refers to the economic value generated through market transactions, measured in terms of profit margins, revenue, or price margins.

For instance, the profit margin gained by assembling components and selling them as finished products represents a typical form of monetary value.

6.4.2.2 Networked Value

This dimension highlights how products or services can connect and integrate with other products, generating extended or networked value. This aspect applies not only to the physical space (e.g., manufacturing and logistics) but also to the virtual space, encompassing data, branding, and digital networks.

For example, a smartphone derives substantial value not just from its hardware components but from how it integrates with apps, cloud services, and social networks, forming a dynamic ecosystem. This connectedness amplifies its value, demonstrating that value is not merely a standalone measure but one that is significantly enhanced through connectivity.

Figure 6.3 illustrates how the Smiling Curve positions Japan (Toyota), Germany (Volkswagen), and the United States (Ford) across the upstream, midstream, and downstream stages of manufacturing. The vertical axis represents economic monetary value, while the horizontal axis reflects the flow of the production process.

However, the key point here is that the value generated by these products is not limited to economic gains (Simon, 1955). The physical existence of manufactured goods also generates social and cultural value, creating a multi-layered structure of value within an organized society.

Figure 6.3 visualizes this hierarchical perspective of value, demonstrating how value is created not only in the physical space (tangible assets), such as manufacturing and assembly, but also in the virtual space (intangible assets), including data, branding, and intellectual property.

Thus, the Smiling Curve serves as a comprehensive framework that shows how products generate value not only in monetary terms but also as multidimensional entities, encompassing economic, social, and cultural dimensions throughout both physical and virtual spaces.

6.4.3 Organizational Structures for SCM: The Japanese "Keiretsu," the German "Konzern," and the Anglo-American Model

From this point forward, we will analyze the manufacturing industries of Japan, Germany, and the United States using a multidimensional perspective that considers both tangible (physical) and intangible (virtual) value structures.

Previous studies have suggested that Japanese and German automakers adopt a producer-oriented perspective, while American automakers are more aligned with a user-oriented perspective in the automotive industry. Based on these studies, this analysis will examine the following hypothesis:

- Producer Perspective (Japan and Germany): Emphasis on tangible (physical) value creation.
- User Perspective (United States): Emphasis on intangible (virtual) value creation.

Next, we will explore how these differing perspectives are reflected in the organizational structures of each country's manufacturing sector. Specifically, we will examine:

- Japan's Keiretsu structure, characterized by vertically integrated supply chain.
- Germany's Konzern structure, emphasizes centralized and bureaucratic control, with strong coordination between parent companies and subsidiaries.
- The shareholder-centric model prevalent in the United States, focusing on financial performance and market-driven strategies.

Through this comparative analysis, we gain a comprehensive understanding of how value-creation strategies differ across regions, aligning with the tangible and intangible dimensions of the Smiling Curve framework.

Figure 6.3 visualizes the distinctive characteristics of Japanese, German, and American manufacturing industries. In this framework, Germany is positioned as a hybrid model, situated between Japan's producer-oriented approach and America's user-oriented focus. However, it is essential to note that each country inherently contains elements of both tangible and intangible value creation, and the typologies presented here are generalized for analytical purposes (Turnbull, 1997). Comparative Analysis of Corporate Structures in Japan, Germany, and the United States Based on Network Models.

The development of stakeholder theory has evolved from focusing solely on the dyadic relationships between companies and individual stakeholders to emphasizing network structures where multiple stakeholders interact and exert influence on each other. In this context, the use of social network analysis becomes essential to assess how these interactions shape organizational behavior.

Rowley extends Freeman's theory by analyzing how various stakeholders in society interact with companies, introducing three distinct models (Rowley, 1997):

- Traditional Management Model: This model centers around the company and its primary stakeholders such as shareholders, suppliers, customers, and employees.
- Stakeholder Model: Expanding on the traditional model, this framework includes broader stakeholders such as environmental organizations and local communities, highlighting a more comprehensive range of relationships.
- Network Model: This model evaluates the influence of stakeholders through two key metrics—density and centrality.

Rowley proposes that the extent of stakeholder influence is determined by the following network attributes:

- Density: The degree of interconnections among stakeholders within the network. High density fosters the sharing of norms and intensifies stakeholder pressure.
- Centrality: The strategic positioning of the organization within the network. High centrality enables the organization to act as a coordinator or gatekeeper, managing the flow of information and resources.

According to Rowley, moving beyond dyadic analysis and focusing on the overall network structure and interdependencies provides a more nuanced understanding of how firms respond to multiple and simultaneous pressures from various stakeholders.

By integrating this network perspective, it becomes possible to better explain how firms strategically navigate external pressures such as technological changes—a central theme of this analysis.

6.4.3.1 The Japanese "Keiretsu"

Japanese Keiretsu refers to a corporate group structure based on cross-shareholding and trading relationships between companies, aiming to strengthen inter-company ties and build long-term stability (Crane & Matten, 2010).

- Vertical Keiretsu: Manufacturers and their suppliers form tight-knit alliances (e.g., Toyota and its suppliers).

- Horizontal Keiretsu: Conglomerates like Mitsubishi, Mitsui & Co., and Sumitomo Corp. hold cross-shareholdings across various industries, thereby diversifying risk.

From a network model perspective, Keiretsu structures exhibit the following characteristics:

- High Density with Moderate Centrality: Within the Keiretsu, a highly dense network is formed, facilitating robust inter-company communication. However, the flow of information to external stakeholders is relatively limited.
- Smooth Information Flow and Central Influence: While information exchange within the network is seamless, central firms (e.g., Toyota) hold significant influence over the decision-making processes of affiliated companies.

6.4.3.2 The German "Konzern"

In Germany, a Konzern refers to a legally recognized corporate group characterized by complex structures involving both managerial and ownership relationships. These structures include not only managerial ties, but also legal parent-subsidiary relationships based on equity ownership and voting rights (Baums, 1994). The strong influence of banks, often lacking a "firewall" between financial and industrial sectors (referred to as "Macht der Banken"), is a defining feature of this structure.

German Corporate Governance System

The German corporate governance system is heavily reliant on labor unions. According to the German Stock Corporation Act (Aktiengesetz), all publicly listed companies (Aktiengesellschaften) are required to adopt a two-tier board structure:

- Management Board (Vorstand): Responsible for day-to-day operations.
- Supervisory Board (Aufsichtsrat): Holds final decision-making authority and has greater influence over workforce restructuring than the management board.

The supervisory board typically consists of 20 members, with ten elected by shareholders and the remaining ten representing employee interests. This structure fosters stability in German manufacturing industries and promotes employee representation in decision-making (Shimizu, 2012; Clark & Fujimoto, 1991).

IG Metall

Volkswagen employs 600,000 people globally, with 120,000 in Germany alone. In comparison, Toyota employs approximately 350,000 workers and General Motors 200,000. German automakers with such vast workforces must negotiate with powerful nationwide labor unions like IG Metall.

IG Metall is Germany's largest industrial union, representing metalworkers, and is instrumental in maintaining high-quality standards, particularly in premium brands like Daimler and BMW. It is the country's, and Europe's, largest industrial union. They are responsible for education systems aimed at maintaining their high-quality products, such as luxury automobiles like those of Daimler and BMW.

- High Density and Strong Centrality: Parent companies, subsidiaries, and labor unions are closely interconnected, resulting in a tightly controlled network structure.
- Strongly Shared Norms: Organizational norms are widely shared, and business decisions are made considering the collective interests of the entire network.

6.4.3.3 The Anglo-American Model

The Anglo-American model is characterized by a governance structure that prioritizes shareholder interests. In this model, social security is often managed through the stock market, making the protection of shareholder assets (particularly pension funds and workers' assets) a primary objective of corporate management. The influence of large institutional investors is a defining feature of this structure.

However, when applying Freeman and Rowley's Stakeholder Network Model, it becomes evident that the heightened focus on shareholder interests can weaken connections with other stakeholders (such as employees, suppliers, and consumers). This can lead to a risk where corporate decision-making becomes increasingly oriented toward short-term profit maximization.

To address this risk, regulatory frameworks in the Anglo-American model require companies to ensure transparency in corporate governance. This includes holding companies accountable not only to shareholders but also to other stakeholders, thereby promoting a broader sense of responsibility.

- **Japanese Model:** The Japanese model is characterized by vertical integration based on the Keiretsu network. Close ties between manufacturers and suppliers foster mutual interdependence, promoting long-term strategic planning rather than short-term profit. Dense networks with moderate centrality enable efficient information sharing and collective decision-making, as seen in companies like Toyota and its supplier network.
- **German Model:** The German model is based on a centralized Konzern structure, where parent companies, subsidiaries, and labor unions are closely interconnected. The supervisory board (Aufsichtsrat) includes representatives from labor unions, ensuring that corporate strategies consider employee and community interests. This structure reinforces long-term stability and strong labor relations, contributing to sustainable growth.
- **Anglo-American Model:** The Anglo-American model emphasizes short-term profit maximization, with rapid technological adoption aimed at increasing

shareholder value. Institutional investors exert significant influence, potentially weakening the company's ties with other stakeholders. The focus on shareholder returns can undermine long-term sustainability, as corporate decisions often prioritize capital efficiency over stakeholder relations.

The shareholder-centric governance structure of the Anglo-American model emphasizes the visibility of short-term capital efficiency but risks eroding relationships with other stakeholders. In contrast, the Japanese and German governance models, with their dense networks and balanced stakeholder representation, aim to reinforce inter-organizational relationships and promote long-term sustainability.

Thus, as technological advancements continue to reshape corporate strategies, a comparative analysis of these governance models becomes increasingly relevant in assessing how companies balance shareholder interests with broader stakeholder engagement (see Table 6.4).

Table 6.4 Characteristics of Japanese, German, and Anglo-American Models

	Upstream	Midstream	Downstream
	Producer's Perspective	Producer/User	User Perspective
	Tangible (physical)	Tangible/Intangible	Intangible (virtual)
	Japan	*Germany*	*USA*
People	**Dense and medium-central structure,** keiretsu in Japan: Vertical keiretsu: Manufacturers and their suppliers work closely together	**Dense and highly centralized**: Each company has a strong control over the parent company "Konzern", subsidiaries, and labor unions.	**Low-density and highly centralized structure:** Each company maintains independence centered on its shareholders, but information is rapidly shared through the stock market.
Things	Parts manufacturing (kaizen)	Product assembly (modular, Industrie 4.0)	Distribution and sales, use of GAFAM and software defined vehicle
Money	Horizontal keiretsu: Multiple companies collaborate in a shareholding relationship	The relationship between the controlling company and the controlled company is strong, and it is tied by shares and voting rights.	Angel investors provide initial capital and influence in exchange for stock ownership
Information	Competitive advantage that conforms to EU regulations but is flexible to U.S.-style information regulations	Strict laws and regulations based on privacy regulations and emphasis on public control of information	Efficient use of information resources with an emphasis on competition

OICA, Author's given

6.5 From an "Embedded System" to an "Internet of Things" Plus "Data and Service"

6.5.1 Embedded Systems and Cyber-Physical Systems (CPS)

Embedded systems are computer systems integrated into machines or devices to perform specific functions. Although general-purpose computers such as personal computers may broadly fall into this category, the term usually refers to specialized control systems designed for particular applications.

The software in embedded systems does not operate independently; it always functions in tandem with the machine or device it is embedded in. In recent years, embedded systems have increasingly been networked through communication technologies such as RFID. This integration marks the evolution into what is now broadly known as the Internet of Things (IoT).

Such advanced interconnectivity and control mechanisms lead to the emergence of Cyber-Physical Systems (CPS). CPS refers to systems where embedded computing devices monitor and control physical processes, usually in real time, by integrating sensor data and communication with actuation.

A further development of CPS is Industrie 4.0, which applies the networked nature of embedded systems to manufacturing sites. It enables real-time, multi-directional monitoring and control of physical objects, whose positions and flows were previously only understood in uncertain and linear terms. It is anticipated that the number of embedded systems worldwide will grow exponentially, reaching trillions in the coming years.

6.5.1.1 Closed Embedded Systems

The starting point for embedded systems lies in closed systems, such as electric kettles or automotive airbags, which contain microcontrollers (microcomputers) that operate independently to execute specific functions (see Fig. 6.4). These systems do not communicate externally and perform internal, self-contained control. Machine tools used in automobile manufacturing also fall under this category and are typically composed of mechatronic components, representing the narrow definition of embedded systems.

6.5.1.2 Networked Embedded Systems

Traditionally, embedded systems communicated in a limited manner via Machine-to-Machine (M2M) interactions. However, advances in semiconductor technologies—guided by Moore's Law—have enabled these systems to become more complex and connect with external networks.

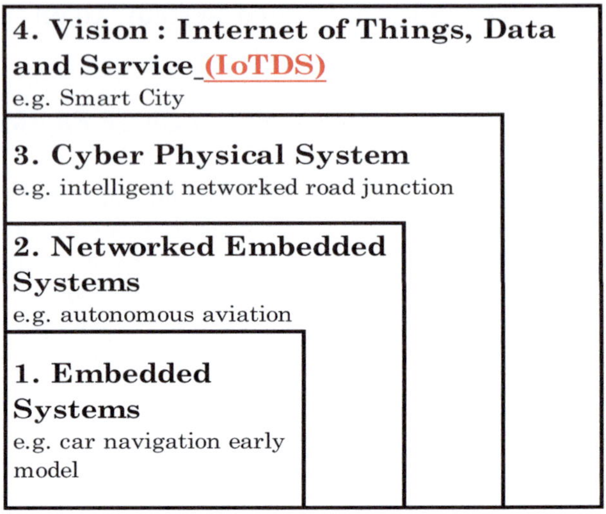

Fig. 6.4 Evolution of embedded systems into the IoT, data, and services (Created by the author)

For example, automated flight systems in aircraft allow each plane to exchange data with neighboring aircraft about speed, altitude, and flight paths. This enables autonomous navigation and operation, showcasing the capabilities of next-generation networked embedded systems.

6.5.1.3 Cyber-Physical Systems (CPS)

Today, with the proliferation of the Internet of Things (IoT), various devices are now connected in real time via the Internet. For example, in traffic information services accessed through smartphones, users' location data is immediately processed in the cloud and used to provide route guidance that reflects current congestion conditions. This system, which integrates the physical world (physical systems) with the digital world (cyberspace) and enables coordinated control, is known as a Cyber-Physical System (CPS).

6.5.1.4 From the Internet of Things to Data and Services

Beyond the standard definition of IoT, the concept of smart interconnectivity has contributed to significant outcomes, such as reducing CO_2 emissions. For instance, in today's smart healthcare systems, patients and doctors are connected through networks that enable remote diagnostics and home-based care.

In fields such as manufacturing, logistics, and transportation, Internet-enabled remote monitoring systems are enabling the development of autonomous production frameworks. These transformations are underpinned by the Industrie 4.0 concept, which replaces the traditional linear flow of production with non-linear, multi-directional, and multi-layered process management.

A key anticipated advancement is self-organization: systems that autonomously assess their operational loads and perform maintenance or repairs at appropriate times. Even if localized issues arise, these systems maintain production without halting the entire operation, supported by flexible backup mechanisms (acatech POSITION PAPER, 2011).

6.5.2 *Usability of IoT and Industrie 4.0*

While mechatronics was the main driver during the eras of Industrie 2.0 and 3.0 (from the 1970s onward), the current waves of Industrie 4.0 and 5.0 place Cyber-Physical Systems (CPS) and the Internet of Things (IoT) at the center of industrial transformation.

In addition, emerging technologies such as next-generation AI, Cloud Computing (CC), Big Data Analytics (BDA), Blockchain Technology (BCT), and the Metaverse are expected to further enhance the intelligence, scalability, adaptability, and sustainability of next-generation mechatronic systems (Zhuming et al., 2025).

Industrie 4.0 is an advanced industrial model that leverages IoT and CPS to realize non-linear and autonomous production management across various sectors including manufacturing, logistics, and healthcare. "Smart Things" are interconnected and cooperate with one another to optimize production lines and supply chains in real time. The vast amount of data collected through IoT is processed and analyzed via BDA to support intelligent decision-making. As a result, the traditional static and reactive systems are being transformed into dynamic and predictive systems (Kagermann et al., 2013).

6.5.3 *Toyota's New Global Architecture (TNGA) & Volkswagen, Modular Transversal Toolkit or Matrix (MQB)*

Germany's Volkswagen (VW) developed a unified production platform known as MQB (Modular Transversal Toolkit or Matrix). This platform allows various car models to share common structures and components, reducing production costs while maintaining product differentiation. In areas where VW lacks expertise, it outsources functions to external suppliers—leading to a flexible, "LEGO-like" assembly model. However, this outsourcing strategy also raises concerns about the "black-boxing" of technological processes, where internal mechanisms become opaque.

Meanwhile, Japan's Toyota introduced TNGA (Toyota New Global Architecture), a design philosophy that aims to enhance production efficiency and flexibility. Inspired in part by Volkswagen's platform strategy, TNGA (now developed to e-TNGA) is Toyota's future-oriented initiative to meet the demands of vehicle electrification, while improving product quality, cost efficiency, and environmental and safety performance. It can also be viewed as a strategic shift away from Toyota's previously "Galápagos-like" isolated production system.

Both VW's and Toyota's production networks are evolving into Cyber-Physical Systems (CPS) that integrate physical systems (the real-world factory) with digital systems (information technology). Moving forward, understanding how CPS continues to connect and expand across factory floors and digital platforms will be crucial for the development of future industrial ecosystems.

6.5.4 Evolution of Automobiles and the Impact of IoT Technology

In recent years, automobiles have evolved from mere mobility machines into "moving information terminals." With the advancement of IoT (Internet of Things) technologies, modern vehicles are equipped with numerous sensors that can collect and analyze diverse and sophisticated data in real time during driving. As a result, not only the functions of individual vehicles but also the structure of the entire transportation system is being reconsidered and reorganized (Porter & Heppelmann, 2014).

Traditionally, automobiles have been based on two key functions: "engine-powered driving" and "human-operated control." Today, while engine performance and miniaturization have significantly advanced, the full realization of automated driving remains under development. Amid these changes, IT companies such as Google are entering the field of autonomous vehicle development. Although they lack experience in automobile manufacturing, they leverage their expertise in the Internet and data analysis to gain a competitive edge in developing technologies that simulate human driving behavior.

In terms of power systems, although electric motors are expected to replace internal combustion engines in the future, approximately 90% of vehicles currently still rely on internal combustion engines. Nevertheless, as these engines and batteries become connected to the Internet, they are beginning to take on a new role as part of the broader social energy infrastructure.

The integration of automotive and IT technologies is significantly affecting industrial structures and corporate governance. Notably, in Germany and Japan, there is a strong emphasis on the "producer perspective," while in the United States, the "consumer perspective" tends to dominate, influencing the direction of technological adoption and corporate strategies.

For example, the application of IoT in automobile production is enabling a departure from conventional linear assembly lines. Instead, bottlenecks can now be detected automatically, and production processes can be flexibly adjusted. Moreover, this shift is not limited to manufacturing processes; the final product—the automobile—is also evolving into a "customizable product" that responds to diverse consumer needs. In the future, we may see production systems in which vehicle platforms autonomously move to where the necessary parts are located and are assembled dynamically, eliminating the traditional concept of fixed production lines.

6.6 Conclusion

6.6.1 Conclusion of the Chapter

This chapter has examined the transformation of technological development and corporate governance systems in the automotive industry, focusing on Germany, Japan, and the United States. The analysis was conducted through six key comparative dimensions.

The matrix presented at the beginning of the chapter serves as a multifaceted analytical framework for understanding the strategic orientations of the manufacturing industries in each country, covering the following perspectives:

1. **Trends in Manufacturing's Share of GDP**: Reflects the stability or growth trajectory of each country's industrial structure.
2. **Industrie 4.0 and Governance Structure**: Highlights modularization, co-determination systems, and shareholder-centric governance models.
3. **IoT Usability Perspective**: Compares "producer-centric" approaches like smart factories with "user-centric" concepts like smart cars.
4. **Smiling Curve & Value Chain Positioning**: Identifies which segments of the value chain each country leverages for high value creation.
5. **Production System & Embedded Technologies**: Examines platform strategies (e.g., TNGA, MQB) and technical system design.
6. **Cyber-Physical Systems (CPS) & ICT Development**: Analyzes integration of AI, sensors, and cloud-based systems as foundations for next-generation industry.

These six dimensions reveal cross-industry structural transformations, illustrating how digitalization and IoT technologies are redefining industrial boundaries and reshaping corporate governance through networked value chains.

6.6.1.1 GDP Trends for Manufacturing Industries

Since the 2000s, Germany has maintained a stable share of over 20% of its GDP in manufacturing. This is supported by high-quality mechanical engineering and an export-oriented industrial structure, making Germany a core manufacturing hub within the EU. Japan, on the other hand, has shown an upward trend—from 33% to 35%—driven by advancements in electrification and hybrid technologies. In contrast, the United States has seen a slight decline from 19% to 18%, as its economy shifts toward ICT, finance, and services.

6.6.1.2 Industrie 4.0 & Governance Structure

Germany features a Konzern-style centralized governance model that combines modular production with institutionalized co-determination, forming a unique structure of corporate governance. In Japan, the dominant Keiretsu system is based on long-term inter-firm cooperation, characterized by dense networks and collaborative transactions. The United States follows the Anglo-American model, where shareholder interests dominate, and market mechanisms with a focus on short-term profit optimization form the core of governance.

6.6.1.3 Viewpoint Relating to IoT Usability

Both Germany and Japan utilize a producer-centric "Smart Factory" model, leveraging IoT to enhance quality control and process efficiency. Germany emphasizes autonomous manufacturing through interconnected machinery. Japan incorporates IoT into the Toyota Production System, particularly in Kaizen (continuous improvement). In contrast, the United States has adopted a user-centric "Smart Car" model, emphasizing UX (user experience) in areas such as autonomous driving and mobility services.

6.6.1.4 Smiling Curve & Value Chain Network

Germany leverages Industrie 4.0 to enhance value in upstream (design, R&D) and midstream (manufacturing) segments, focusing on high value creation at the front end of the value chain. Japan maintains competitiveness in the midstream segment, balancing cost and quality through hybrid and EV technologies. The United States specializes in downstream sectors (sales and services), capitalizing on its strengths in software and digital services.

6.6.1.5 Production System & Embedded System

Volkswagen's MQB (Modular Transversal Toolkit) enables cost reduction and flexible production through modularization across multiple vehicle types. Toyota's TNGA (Toyota New Global Architecture) integrates process optimization with on-site improvement, aiming for efficient yet adaptable manufacturing. US manufacturers emphasize software-defined vehicles (SDVs), leveraging IoT-based control systems and cloud integration for product differentiation.

6.6.1.6 Cyber-Physical System & ICT-Related Development

Germany promotes a privacy-conscious and sustainable digital transformation under frameworks like GDPR. Japan's Society 5.0 vision aims to build a holistic CPS (Cyber-Physical System) integrating AI and IoT. In the United States, EV manufacturers such as Tesla collaborate with GAFAM platforms to deliver AI-, cloud-, and big data-driven services centered on user experience.

Crossing Industrie Boundaries

The six perspectives explored in this chapter—(1) GDP trends in manufacturing, (2) governance structures and Industrie 4.0, (3) IoT usability perspectives, (4) Smiling Curve and value chain analysis, (5) production systems and embedded technologies, and (6) CPS and ICT evolution—not only transform internal industrial structures but also blur the boundaries between traditional industries.

Especially with the rise of value chain specialization and digital networking, the once-distinct boundaries between the automotive, energy, and ICT sectors are becoming increasingly fluid. This trend is most apparent in Mobility as a Service (MaaS), connected cars, and energy management systems. We are witnessing a shift from an era of industry-specific value creation to one of inter-industry value co-creation through collaborative networks (BlackRock, 2017).

One of the most noteworthy developments is the increasing convergence between the automotive and energy industries. While the proliferation of electric vehicles (EVs) is being promoted from the perspective of environmental sustainability and the circular economy, it also extends beyond these frameworks. Through interaction with power grids, new energy utilization models such as Vehicle-to-Home (V2H) and Vehicle-to-Grid (V2G) are emerging, enabling vehicle-mounted batteries to supply electricity to homes and the grid (see in Fig. 6.5) (acatech POSITION PAPER, 2011).

In essence, the shift to EVs represents a conceptual transformation from "vehicle" to "mobile energy source." Automotive manufacturers such as Toyota, Volkswagen, and Tesla are now evolving into key actors in a new cross-industry ecosystem, connecting with renewable energy systems and smart grids.

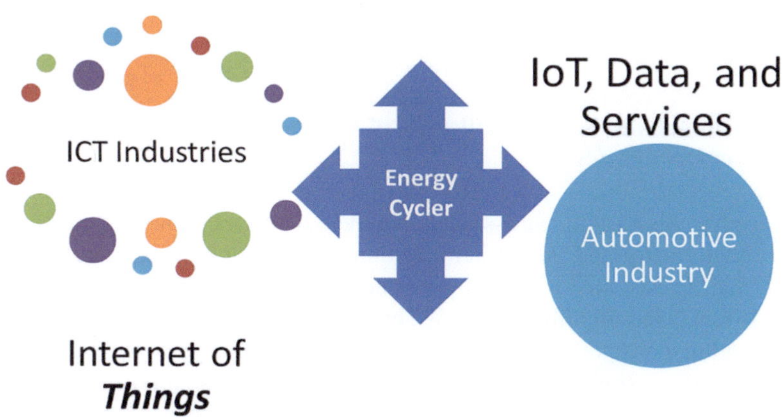

Fig. 6.5 Meaning of "IoT, *Data and Services*" (Created by the author)

By analyzing the six dimensions discussed earlier, it becomes clear that modern manufacturing is no longer limited to "making things." Rather, it is evolving into "creating value" through integration with information, energy, and social systems.

6.6.1.7 Blockchain Governance

In today's industrial landscape, the integration of technological innovation and governance is a critical issue in the pursuit of a sustainable society. In the domain of blockchain governance, system stability and security are indispensable, and the question arises as to which data should be utilized from the vast amounts being generated in order to build a safe and resilient social infrastructure.

Meanwhile, in the automotive sector, advancements in fuel efficiency and embedded systems are opening new paths toward sustainability. Two key elements define this evolution:

- The shift in energy sources (electrification).
- The advancement of control systems, including IoT-based smart systems.

Modern vehicles incorporate complex embedded systems to control functions such as fuel injection and driving stability. These systems are increasingly being interconnected via IoT, leading to greater automation and intelligence.

Furthermore, with the rise of IoT and smart technologies, vehicles are being redefined from mere "transportation devices" to energy storage and distribution units, exemplified by V2G models. This transformation is accelerating innovation through convergence, as the automotive industry crosses traditional boundaries and merges with the energy sector to create new forms of added value.

This structural transformation also resonates with the cybernetic perspective of Stafford Beer, who emphasized that "we should not aim to predict the future, but to understand the present with precision in order to shape future possibilities." His philosophy offers crucial guidance for redesigning technology governance and formulating sustainable policies in the face of growing societal and industrial complexity (Beer, 1973).

6.6.1.8 Ethical and Social Challenges of Digitalization and Its Impact on Democracy

The preceding discussion has assumed that the advancement of digitalization in the automotive industry and the implementation of IoT technologies can be accepted without major resistance by employees and stakeholders involved. However, promoting IoT, AI, and digitalization in the era of big data requires a clear understanding of—and response to—a range of complex challenges.

First, the processing and generation of data demand substantial energy consumption, placing a significant burden on the natural environment. Furthermore, it is essential to support the retraining and upskilling of employees to adapt to these technological changes. Privacy protection of personal data, consideration of disparities in access to data (known as the "Matthew effect"), concerns over the rise of surveillance societies, and the implementation of countermeasures are also crucial ethical and social issues.

Most critically, big data–driven AI and other applications pose the risk of circumventing or even violating existing legal and regulatory frameworks. Already, phenomena such as filter bubbles and echo chambers have become apparent, threatening the foundation of democracy by distorting free decision-making and fair electoral processes. In fact, we have witnessed cases where voter behavior, shaped by biased information environments, has led to political polarization and extreme outcomes.

Addressing these issues requires not only social understanding, public discussion, and institutional responses but also an awareness on the part of individual users. In short, a healthy and sustainable digital society depends on each user's conscious engagement with the ethical and societal implications of the technologies they use (David, 1985).

6.7 Review Questions

1. Relationship between IoT and CPS

   Explain the relationship between the Internet of Things (IoT) and Cyber-Physical Systems (CPS).

   How do these technologies enhance autonomy and real-time responsiveness in modern manufacturing industries?
2. The Smiling Curve and Value Creation in the Automotive Industry

   Using the concept of the Smiling Curve, explain how Japan, Germany, and the United States each create value within their automotive industries.

   In your answer, discuss how differences in industrial structure and governance affect value creation along the value chain.
3. Smart Factory vs. Smart Car

   Describe the difference between a Smart Factory and a Smart Car, and explain how each represents a distinct industrial strategy.

   How do these two models reflect the contrasting producer-centric and user-centric approaches to technological innovation?
4. Network Theory and Stakeholder Relationships

   Using Rowley's (1997) Network Theory, explain how relationships between firms and stakeholders are structured.

   Discuss how the concepts of density and centrality influence the flow of information and power within stakeholder networks.
5. From "Making Things" to "Creating Value"

   In the conclusion of this chapter, the shift from "making things" to "creating value" is emphasized.

   Explain what this transformation means in the context of industrial evolution, particularly with regard to the integration of energy, information, and social systems.

References

acatech POSITION PAPER. (2011). Cyber-physical systems: Driving force for innovation in mobility, health, energy and production. Abgerufen am 12. 9 2017 von http://www.acatech.de/fileadmin/user_upload/Baumstruktur_nach_Website/Acatech/root/de/Publikationen/Stellungnahmen/acatech_POS

acatech POSITION PAPER. (2012). Future energy grid; Information and communication technology for the way towards a sustainable and economical energy system. Abgerufen am 12. 9 2017 von http://www.acatech.de/fileadmin/user_upload/Baumstruktur_nach_Website/Acatech/root/de/Publika

Association (JAMA) Automotive Manufacturers Japan. (2025). JAMA. 参照先. https://www.jama.or.jp/

Baums, T. (1994). *Corporate governance in Germany—System and recent developments*. University of Frankfurt.

Beer, S. (1973). *Desigining freedom.* Series of Massey Lectures. Retrieved October 6, 2017, from http://ada.evergreen.edu/~arunc/texts/cybernetics/beer/book.pdf

BlackRock (2017). PUBLIC POLICY; Common ownership. Retrieved from https://www.blackrock.com/corporate/literature/whitepaper/viewpoint-index-investing-and-common-ownership-theories-eng-march.pdf

CECIMO. (2025). *European Association of the Machine Tool Industries and related Manufacturing Technologies*. Retrieved from https://www.cecimo.eu/

Clark, K. B., & Fujimoto, T. (1991). *Product development performance*. Harvard Business School Press.

Crane, A., & Matten, D. (2010). *Business ethics, managing corporate citizenship and sustainability in the age of globalization* (4th ed.). OUP Oxford.

David, P. A. (1985). Clio and the economics of QWERTY. *The American Economic Review*. Retrieved October 6, 2017, from https://econ.ucsb.edu/~tedb/Courses/Ec100C/DavidQwerty.pdf

Executive Office of the President, President's Council of Advisors on Science and Technology. (2014). Report to the President, accelerating U.S. advanced manufacturing. President's Council of Advisors on Science and Technology, p. 4. Retrieved October 6, 2017, from https://www.manufacturingusa.com/sites/prod/files/amp20_report_final.pdf

Galbraith, J. K. (1967). The new industrial state. Princeton University Press. Retrieved October 7, 2017, from https://books.google.co.jp/books?id=8l2G-C8H8IoC&printsec=frontcover&hl=ja&source=gbs_ge_summary_r&cad=0#v=onepage&q&f=false

Kagermann, H., Wahlster, W., & Helbig, J. (2013). Recommendations for implementing the strategic initiative INDUSTRIE 4.0. (Forschungsunion, & a. –N. Engineer, Eds.)

Kurzweil, R. (2010). *The singularity is near*. Duckworth Overlook.

National Accounts Main Aggregates Database. (2026). Comparison of major countries of value added by manufacturing activity, percentage distribution (shares) in GDP. https://unstats.un.org/unsd/snaama.

Organisation Internationale des Constructeurs d'Automobiles (OICA). (2025). Retrieved from https://www.oica.net/

Porter, M. E. (1998). *Competitive advantage: Creating and sustaining superior performance* (1st ed.). Free Press.

Porter, M. E., & Heppelmann, J. E. (2014). How smart, connected products are transforming competition. *Harvard Business Review*. Retrieved October 6, 2017, from https://hbr.org/2014/11/how-smart-connected-products-are-transforming-competition

Rowley, T. J. (1997). Moving beyond dyadic ties: A network theory of stakeholder influences. *Academy of Management Review, 22*(4). https://doi.org/10.5465/amr.1997.9711022107

Shih, S. (2002). *Growing global: A corporate vision masterclass*. John Wiley & Sons.

Shimizu, K. (2012). German heavy-weight management: A case study of Volkswagen. *Proceeding of the APCIM 2012.*

Shimizu, K. (2016). Smart factory or smart car? The concept of IoT usability. *The fifteenth international conference on networks (ICN 2016)* (pp. 68–71). Lisbon, Portugal: IARIA. Retrieved October 7, 2017, from http://www.thinkmind.org/index.php?view=article&articleid=icn_2016_4_10_30098

Shimizu, K. (2017). Technological development in automotive industry and transformation in corporate governance system. *Journal of Governance & Regulation, 6*(3). https://doi.org/10.22495/jgr_v6_i3_p5

Shintaku, J. (2016). Development of offshore production and its impact on domestic manufacturing. *Global Business Research Center: Akamon Management Review*. https://doi.org/10.14955/amr.150501

Siemieniuch, C. E., Sinclair, M. A., & Henshaw, M. (2015). Global drivers, sustainable manufacturing and systems ergonomics. *Applied Ergonomics*. Retrieved October 6, 2017, from http://isiarticles.com/bundles/Article/pre/pdf/43407.pdf

Simon, H. A. (1955). A behavioral model of rational choice. *The Quarterly Journal of Economics, 69*(1). Retrieved October 7, 2017, from https://iiif.library.cmu.edu/file/Simon_box00063_fld04838_bdl0001_doc0001/Simon_box00063_fld04838_bdl0001_doc0001.pdf

The Government of Japan. (2017). *ABENOMICS.* (T. g. Japan, Ed.) Retrieved October 7, 2017, from https://www.japan.go.jp/abenomics/_userdata/abenomics/pdf/170630_abenomics.pdf

Turnbull, S. (1997). Stakeholder governance: A cybernetic and property rights analysis. *Corporate Governance: An International Review, 5*(1).

World Bank. (2025, April). *Manufacturing, value added (% of GDP)—North America, Japan, Germany, China.* Retrieved from https://data.worldbank.org/indicator/NV.IND.MANF.ZS?end=2023&locations=XU-JP-DE-CN&start=1991&view=chart

Zhuming, B., Chris, Z. W., Jianning, C., & Chaomin, L. (2025). Internet of Things (IoT) for sustainable mechatronic systems and cyber-physical systems (CPS). *ICINT2025*.

Part IV
Blockchain Technology

Chapter 7
Blockchain and Biometrics Authorization: What We Actually Count Truly Counts?

Abstract The purpose of this paper is to explore the possibility of sharing value among complexly related stakeholders using blockchain, the core technology of cryptocurrency (Bitcoin) [Nakamoto (*Bitcoin: A peer-to-peer electronic cash system*, 2008)]. In particular, the value interaction between human input (smell, taste, touch, sound, sight) and human output (tactile, symbol, writing, sound, voice) as IoB (Internet of Body) via the blockchain is concerned. This interaction is related to the cognitive circuits in the brain [Turnbull (*A sustainable future for corporate governance theory and practice*, 2012)]. The dilemma is the autonomous decentralization of systems through P2P (peer-to-peer authorization) and the digital invasion of biological privacy. IoB privacy data are a human rights issue and must be protected appropriately and evaluated. Personal physical data give possibilities for understanding effective predictions of the collective human action. This is similer to adopting process of bitcoin into society. In other words, *it is similar to finding a causal correlation with a logical way of thinking as the background of data.* Does society (collective cognition) admit new things like Bitcoin as a payment method? Here, the correlation between e-commerce data, Bitcoin prices, NASDAQ, and other variables is concerned. As a result, Bitcoin strongly correlated with NASDAQ (0.844), the regression coefficient with e-commerce data was negative 5%, significantly negative to the Bitcoin price, and 1% strongly positive with the NASDAQ index (equation: $Y_1 = -6.59 \times 10^{-16} - 0.211X_1^{*} + 0.068X_2 - 0.233X_3 + 0.863X_4^{***}$). It assumed Bitcoin does not work as a major payment method for e-commerce when the price of Bitcoin continues to rises and fluctuates. However, if Tesla and Visa purchase Bitcoin as their payment method, the price will rise due to increased trust. Therefore, this phenomenon shows that the blockchain technology, as a proxy variable, and the NASDAQ, has been very promising.

Keywords Blockchain · Peer to peer (P2P) · Biometrics authorization · Decision-making process · Internet of Body (IoB)

K. Shimizu, *ESG Investment, German Industrie 4.0, and Blockchain*, https://doi.org/10.1007/978-981-95-6927-4_7

7.1 Blockchain

There is legal currency (money) as a method of measuring value. Before the Nixon shock, dollars were proportional to the value of gold (gold standard). As a primary usage, money mediates the exchange of goods and services. However, in the recent Covid19 situation, the value gap between the real economy and the monetary economy is remarkable due to a considerable amount of money supplies from central banks (Piketty, 2014). Gold stands at roughly $10 trillion in 2020 December. For example, how long it took a company to be trillion dollars monetary network; Microsoft 44 years, Amazon 24 years, Apple and Google 22 years, Bitcoin 12 years (Michael, 2021). ICOs (Initial Coin Offerings) have emerged in recent years as a new model of funding business venture, in alternative to the traditional investor-mediated capital markets (e.g., Initial Public Offerings (IPOs) and venture capitals).

There are two essential principles in Blockchain. One is the P2P authentication system, and the other is the computational processing speed that depends on the development of technology such as CPU and GPU (graphics processing unit) development. To understand the P2P authentication system, Fig. 7.1 below shows the P2P method and the client–server model.

Currently, money is a settlement function for mediating various values. Therefore, it can be seen that the function of mediation, i.e., using blockchain, such as Bitcoin, can be used for multiple value exchanges. For example, on YouTube, advertising revenue is determined by the number of followers.

The P2P model is an autonomous decentralized system, and the client–server model is a centralized system. The P2P model shows DAO (Decentralized Autonomous Organization), which realizes democracy in virtual space, see above Fig. 7.1.

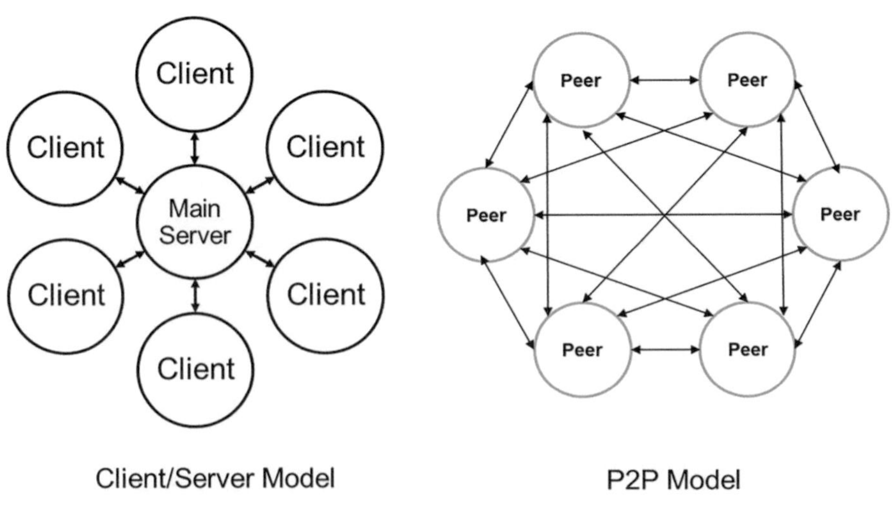

Fig. 7.1 Client/server model versus P2P model. Source: Created by the author with reference to Jaime Galán-Jiménez and Alfonso Gazo-Cervero

From this, it can be understood from the comparison that the P2P model is a stable system in which information is not concentrated. A definition of P2P networking is a set of technologies that enable the direct exchange of services or data between computers. Implicit in this definition is the fundamental principle that peers are equals. P2P systems emphasize sharing among these equals. A pure peer-to-peer system runs without any centralized control or hierarchical organization. Cennamo suggests the success of digital currencies depends on their business type (i.e. platform) and on their technology type (i.e., coin and token) (Cennamo et al., 2020).

The consensus mechanism that Bitcoin uses is called Proof of Work (PoW). PoW is necessary for security, which prevents fraud, which enables trust. This security ensures that independent data processors (miners) can't lie about a transaction. Blocks are produced roughly every 10 min and are made up of transactions. Therefore, the miners (block creators) need powerful computer equipment like a GPU or, more realistically, an application-specific integrated circuit (ASIC).

To get a sense of how much computing power is involved, when Bitcoin launched in 2009 the initial difficulty level was one. As of 2019, it is more than 13 trillion (Euny, 2020).

7.2 Biometrics

What is the boundary between the Internet of Things (IoT) and IoB? In everyday life, consumers spend about 5 h a day on mobile devices and apps such as Facebook, Twitter, and YouTube. The concept of "extended self" has been pointed out by many philosophers. What is a digital identity? We are digitally converted as data via the device. The most recognized example of an Internet of Bodies (IoB) is a pacemaker, a small device placed in the chest to help patients with heart conditions control abnormal heart rhythms with electrical impulses. Increasingly, the external environment of human beings is controlled by smart devices. At the same time, the control from the inside physical body of human beings is added, such as IoB, and the decision-making is greatly affected. The promotion of a bioeconomy that contributes to the Paris Agreement and the SDGs, sustainable economic growth, and the resolution of social issues is positioned as a national strategy in major countries, and ESG investment by institutional investors is also expanding (European Commission, 2012).

The human brain is a massively parallel computer processing many bits of data at once rather than one bit at a time. The personal computer could only "emulate about a million neuron connection calculations per second, which is more than a billion times slower than the human brain" [Kurzweil, 2016]. Figure 7.2 shows the influential factors of human decision-making from input and output as physical actions. When such individual physical activities can be captured as digital data, the interaction between individual autonomy and an organization also could be controlled.

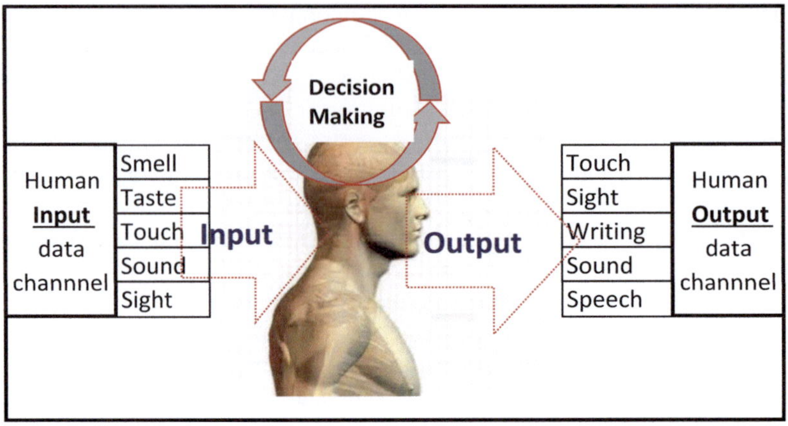

Fig. 7.2 Concept of human transaction. Source: Created by the author (Shimizu, 2016)

Physiological measurements are made possible, for example, through wearable devices supported by widely used and inexpensive sensors. This technology is currently being introduced to markets. Stream mileage, calories burned, exercise intensity, heart rate, and other data to the wearer's mobile device. The Internet of Things (IoT) is built on strong Internet infrastructure. Radio Frequency Identification embedded (RFID-embedded) smart cards are ubiquitous, used for many things, including student I.D. cards, transportation cards, bank cards, prepaid cards, and citizenship cards. The development of biometric identification technologies has continued to advance (Moses et al., 2020). For example; Blockchain methodologies have been successfully adopted in the library, where based on integrated technologies using blockchain and finger-vein biometrics (Meng-Hsuan, 2020 September).

Blockchain can offer an autonomous decentralized privacy protection system explained above. Einstein quote, however "Not everything that can be counted counts, and not everything that counts can be counted."

7.3 Discussion

This paper tried to examine the possibility the value between stakeholders can be shared using blockchain. Significantly, the value interaction between human input and output as IoB via the blockchain was investigated. The dilemma is the autonomous decentralization of systems through P2P and the digital invasion of biological privacy. Murata suggests that in a society where there is a widespread use of artificial intelligence (AI)-based information systems, the truths would become meaningless or worthless. To prevent the emergence of the post-truth society, everyone has to acquire sufficient knowledge and skill for good computing practices, particularly the ability to consider socially and ethically, through undergoing well-organized ICT educational programs (Yamazaki et al., 2020). Only human beings can give meaning to "nothing." Does the layered structure provided by platformers in the digital society represent the "meaning" and "value" that humans give? No matter how much you think for, machines can give nothing for this structure (Norbert, 1950). IoB is trying to visualize the part that functions autonomously by our biological data, especially by the autonomic nerves. However, as Murata finally points out, since we give meaning to and understand the numerical values there, we have to give essential meanings that can be seen there by ethical human beings.

7.4 Is Blockchain Trustworthy?

There are a lot of works to be done for establishing cryptocurrencies, for example, dealing with "Double Spending." Bitcoin can be spent twice as two separate transactions. And there is also "the darknet" side, including illegal porn, settlement for barely legal online gambling, hacking services, and even murder for hire. The anonymous nature of the darknet side facilitates the use of this newly developed cryptocurrency (Stephen, 2015). Many problems have to be solved here.

IoB privacy data are a human rights issue and must be protected appropriately and evaluated. Of course, the digital data obtained by IoB can be difficult to identify individuals and track the behavior of all humans. Still, by clarifying this criterion, it is best for critical personal physical data. It is becoming possible to understand moods and emotions through the regression rates of several variables that contribute to specific behaviors from the movements of the human body using AI. This is considered to effectively predict the future of collective human action. Today, personal data is collected as if it were valueless by many for-profit companies such as GAFAM. However, crypto-assets such as Bitcoin can provide a positive or negative solution for this particular situation. Therefore, many for-profit companies, Visa, PayPal, and Tesla, create network effects from new cryptocurrency business models such as PayPal's peer-to-peer wallet CashApp. Apple is also using smartphones for Apple Card fintech with major securities firm Goldman. From similar movements, physical privacy data is traded by crypto-assets, for example, to provide valuable physical reaction data to pharmaceutical companies.

Various problems of Bitcoin using such blockchain technology will be solved through the trust of society (Shalini, 2020). Bitcoin adapts the mainstreamer adoption of cryptocurrencies by PayPal, Square, VISA, and Tesla. Here we consider the relationship between Bitcoin and the technology companies that should give Bitcoin trust. In other words, the correlation between NASDAQ, an indicator of mainstream tech companies, and Bitcoin prices. As a result, Bitcoin is strongly correlated with NASDAQ (0.844**), the regression coefficient with e-commerce data was negative at 5%, significantly negative to the Bitcoin price, and 0% was strongly positive with the NASDAQ index. The numerical data -target period was from October 2014 to April 2021. Here, we analyze the correlation and regression results between monthly e-commerce transaction data (Y; e-commerce Z) and four variables, such as Bitcoin price (X_1: Price Z), the US interest rate (X_2; U.S. Z)), gold price (X_3; Gold Z), and NASDAQ price (X_4; NASDAQ Z). The monthly e-commerce transaction data (Y; e-commerce Z) use the statistical figures of Euromonitor International (BIDITEX Exchange, 2020) (Euromonitor International, 2021). The variable numbers are calculated using the Z-score to improve the correlation and regression rate analysis accuracy. Therefore, "Z" is written in the table for all variables (see below Tables 7.1 and 7.2).

As a result of Table 7.1 Correlation rate of four variables, e-commerce data is strongly correlated with NASDAQ (X_4; 0.942), and the US interest rate (X_2) correlates negatively. As a result of Table 7.2, the regression coefficient with the Bitcoin price is negative 5% significantly, and 0% strongly positive with the NASDAQ index. See Fig. 7.3, from the e-commerce and Bitcoin price-scatter plots, it shows an unusual value, so it shows a 5% significant, negative value. It could assume Bitcoin does not work as an alternative payment method for e-Commerce when the price of Bitcoin continues to rise and fluctuates. However, many major fintech companies, Visa, PayPal, and Amazon, seem like creating network effects from these cryptocurrency business models, the price will rise as trustworthy. Therefore, this phenomenon shows that the blockchain technology as a proxy variable for the NASDAQ has been very promising. Also, tech companies were usually inversely correlated with euro and dollar interest rates. This is because government bond

Table 7.1 correlation table of four variables

	Price.Z	US.Z	Gold.Z	NASDAQ.Z	e-commerce.Z
Price.Z	1				
US.Z	−0.370	1			
Gold.Z	0.654	−0.791	1		
NASDAQ.Z	0.844	−0.499	0.884	1	
e-commerce.Z	0.717	−0.508	0.877	0.942	1

(Created by the author)

Table 7.2 Multiple regression table of four variables

	Estimate	Std. Error	t value	Pr(>|t|)
(intercept)	−6.59E−15	0.034	−1.92E−13	1.000
'Price Z'	−0.211	0.076	−2.761	0.007*
'US Z'	0.068	0.087	0.781	0.438
'Gold Z'	0.223	0.172	1.300	0.198
'NASDAQ Z'	0.956	0.163	5.862	1.147E-07***

Signif. codes: 0.001 "***" 0.01 "**" 0.05 "*"

Equation: $Y_1 = -6.59\times10^{-16} - 0.211X_1^{*} + 0.068X_2 - 0.233X_3 + 0.863X_4^{***}$

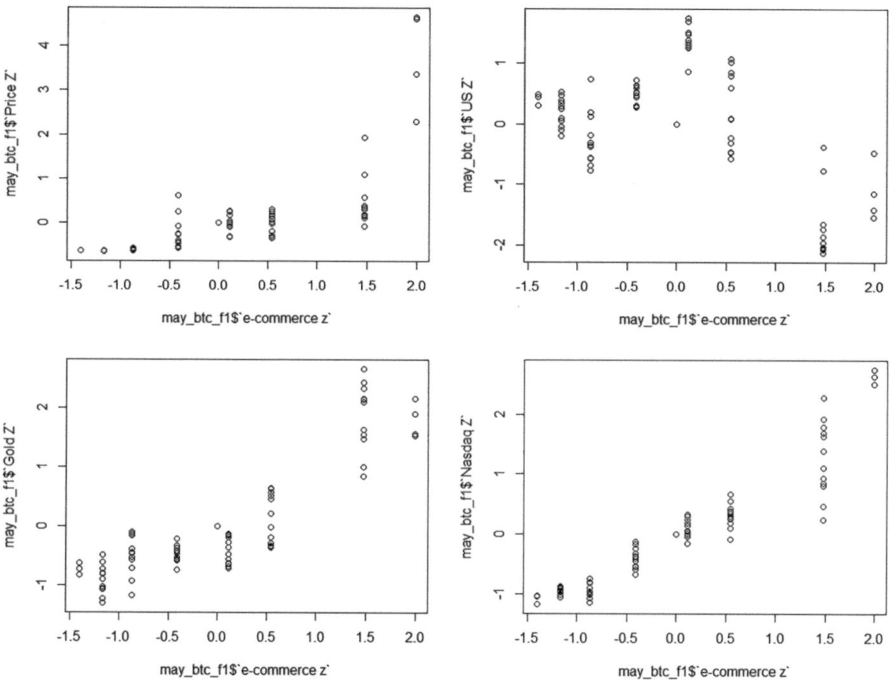

Fig. 7.3 Scatterplot of four variables. Source: Created by the author, using R

yields are preferred over technical investment for certainty during rising interest rates (see all Fig. 7.3 Scatter plot of four variables).

7.5 Review Questions

1. Q1. Explain the difference between the Peer-to-Peer (P2P) model and the Client–Server model in blockchain systems.
   How does the P2P system contribute to decentralization and data security?
2. Q2. According to the regression analysis in the text, Bitcoin's price has a significant negative relationship with e-commerce transaction data.
   Explain why Bitcoin is considered unstable as a major payment method for e-commerce.
3. Q3. What is the purpose of the Proof of Work (PoW) mechanism in blockchain technology?
   Explain how PoW contributes to the reliability and trust of the Bitcoin network.
4. Q4. The text shows a strong positive correlation (r = 0.844) between Bitcoin prices and the NASDAQ index.

What does this relationship imply about the connection between cryptocurrency markets and technology stocks?

5. Q5. Why does the author regard IoB (Internet of Body) data as a human rights issue?

   Discuss the potential ethical and legal challenges of collecting and using biological data in blockchain systems.

6. Q6. Based on the regression model shown in the text:

$$Y_1 = -6.59\times10^{-16} - 0.211X_1 + 0.068X_2 - 0.233X_3 + 0.863X_4$$

where;
Y_1: e-commerce transaction volume (Z-score)
X_1: Bitcoin Price (Z-score)
X_2: US Interest Rate (Z-score)
X_3: Gold Price (Z-score)
X_4: NASDAQ Index (Z-score)
Answer the following:

(a) Which variable has the strongest positive influence on e-commerce data?
(b) Which variable negatively affects it most significantly?
(c) Interpret what the negative coefficient for Bitcoin Price (−0.211) suggests about its role in online payment behavior.
(d) Suppose NASDAQ rises by one standard deviation. Estimate the expected change in e-commerce transactions based on the model.

References

BIDITEX Exchange. (2020, January 22). *7 Countries with the most bitcoin hodlers*. Retrieved from https://medium.com/@biditex/7-countries-with-the-most-bitcoin-hodlers-503b205d926f

Cennamo, C., Marchesi, C., & Meyer, T. (2020). Two sides of the same coin? Decentralized versus proprietary blockchains and the performance of digital currencies. *Academy of Management Discoveries, 6*(3). https://doi.org/10.5465/amd.2019.0044

Euny, H. (2020, November 18). *Investopedia*. Retrieved from How does bitcoin mining work? https://www.investopedia.com/tech/how-does-bitcoin-mining-work/

Euromonitor International. (2021). *Passport*. Retrieved from https://www.portal.euromonitor.com/portal/magazine/homemain

European Commission. (2012). *Innovating for sustainable growth: A bioeconomy for Europe*. Retrieved from https://ec.europa.eu/research/bioeconomy/

Jaime, G.-J., & Alfonso, G.-C. (2011). Overview and challenges of overlay networks: A survey. *International Journal of Computer Science & Engineering Survey*. Retrieved from https://www.researchgate.net/publication/50199321_Overview_and_Challenges_of_Overlay_Networks_A_Survey

Kurzweil, R. (2016). *The singularity is near: When humans transcend biology* (Japanese Ed.). NHK Publishing.

Meng-Hsuan, F. (2020 September). Integrated technologies of blockchain and biometrics based on wireless sensor network for library. *Information Technology and Libraries*. https://doi.org/10.6017/ital.v39i3.11883

Michael, S. (2021). Can bitcoin be replaced or fail? Retrieved from https://www.youtube.com/watch?v=LdwSxwl9BQk

Moses, A. A., Na, C., Jeng-Shyang, P., Hong-Mei, Y., & Bin, Y. (2020). Securing fingerprint template using blockchain and distributed storage system. *MDPI*. https://doi.org/10.3390/sym12060951

Nakamoto, S. (2008). *Bitcoin: A peer-to-peer electronic cash system*. Bitcoin.org. Retrieved from https://bitcoin.org/ja/bitcoin-paper

Norbert, W. (1950). *The human use of human beings; Sybernetics and society*. Retrieved from http://21stcenturywiener.org/wp-content/uploads/2013/11/The-Human-Use-of-Human-Beings-by-N.-Wiener.pdf/2014.05.10

Piketty, T. (2014). *Capital in the twenty-first century*. Belknap Press World. Retrieved from https://dowbor.org/blog/wp-content/uploads/2014/06/14Thomas-Piketty.pdf

Shalini, N. (2020, December 19). *Market insider*. Retrieved from Bitcoin's market cap could hit $1 trillion in 2021 as its growing reserve currency status drives adoption higher, a cryptocurrency expert says: https://markets.businessinsider.com/currencies/news/bitcoin-market-cap-could-hit-1-trillion-2021-crypto-expert-2020-12-1029908224#:~:text=Bitcoin%20has%20a%20current%20market,stands%20at%20roughly%20%2410%20trillion

Shimizu, K. (2016). Socio-cybernetic approach into the triumvirate: Stakeholder governance between management, shareholders and employees. *An Enterprise Odyssey. International Conference Proceedings*, 273–280. Retrieved from https://search.proquest.com/docview/1815362010?pq-origsite=gscholar&fromopenview=true

Stephen, S. (2015). *Bitcoin: The Napster of currency* (pp. 581–641). J.D. University of Houston Law Center. Retrieved from http://www.hjil.org/articles/hjil-37-2-small.pdf

Turnbull, S. (2012). *A sustainable future for corporate governance theory and practice*. Principal: International Institute for Self-governance. Retrieved from https://papers.ssrn.com/sol3/papers.cfm?abstract_id=1987305

Yamazaki, T., Murata, K., Orito, Y., & Shimizu, K. (2020). *Post-truth society: The AI-driven society where no one is responsible*. Universidad de La Rioja: ETHICOMP 2020. Obtenido de https://www.researchgate.net/publication/343099318_Post-Truth_Society_The_AI-driven_Society_Where_No_One_Is_Responsible

Chapter 8
Bullshit Blockchain Mining?

Abstract Smartphones and high-speed computers h3ave advanced our productivity sufficiently and are vital to our daily lives. However, our everyday environment is not much different from before and is surrounded by more "bullshit jobs." It's as if someone were out there making up bullshit jobs to keep us all working. Here, let us consider the meaning of "bullshit." Indeed, Bitcoin has been touted as an innovative method of information security as distributed ledger technology (DLT). GAFA's "centralization of information" can be decentralized by this DLT technology. However, the spread of blockchain technology is considered "bullshit" for the following three reasons. First of all, the issue of electricity consumption in blockchain mining and the industry as a whole, and secondly, whether the blockchain technology is adaptable to the organizational structure and what is the bureaucratic "bullshit job"? Finally, there are some applicable fields in blockchain technology and their limitations.

Blockchain mining cannot wipe out uncertainties about the future. Still, breaking through the current situation, blockchain mining will be a ray of light in the darkness for a globally distributed economy. To try to automate "bullshit jobs" through human enhancement such as the Internet of Ability (IoA), it is necessary to carefully analyze, break down "bullshit jobs," and render unto man's jobs.

Keywords Blockchain governance · corporate and stakeholder governance · Internet of Ability (IoA) · Institutional crypto-economics

K. Shimizu, *ESG Investment, German Industrie 4.0, and Blockchain*,
https://doi.org/10.1007/978-981-95-6927-4_8

8.1 Introduction

Satoshi, 32 years old, has been doing Ethereum mining since 2020 as a side business alongside his main business.[1] He downloaded the software, tried mining, and got a few dollars in terms of the conversion rate of Ethereum for the first time. He started thinking that it would work somehow. He built a specialized P.C. for mining, bought a video card RXT 2080 Ti from Nvidia, and started mining. However, the P.C. dedicated to mining did not work correctly and was initially challenging. After all, mining heats his PC about 70 degrees, and it needs to be released with air-conditioner. His total revenue will be about $70, excluding the electricity bill spending of about $30 per month. This means he can then make about $40 profit per month.[2] Of course, there are possibilities where the data is rejected in the mining pool.[3] This story is an individual-based story of Satoshi.

Thus, mining is not heavy labor using limbs like a gold rush or a coal mine worker. Mining is simply setting up a computer for crypto-assets[4] and letting the computer machine do calculations 24 h a day and 365 days a year, using electricity constantly. This does not involve any physical labor work. Can this be called "Mining"? It is a kind of typical "Bullshit job."

The rest of the paper is structured as follows. Chapter 2 addresses the theoretical background of what a "bullshit job" is in a bureaucratic organization. Chapter 3 investigates the power utilization of mining, an incentive for blockchain. Finally, we will present a new way to use blockchain for management. Next, I would like to explain the definition of "Bullshit job."

---

[1] Two types of mining methods. There are two main types of mining: ASICs and GPUs. Initially, an ASIC, a device designed specifically for mining, can dig Bitcoin (BTC). And with GPU mining using a graphics board, you can earn Ethereum and Altcoin. Due to the peculiarities of ASIC devices, Bitcoin miners are oligopolies by several companies worldwide.

[2] Calculated by https://www.nicehash.com/profitability-calculator/nvidia-rtx-2080-ti

[3] Looking at individual mining work, there is a "mining pool" method. Mining is becoming increasingly popular with high-speed computer devices compatible with home computers in recent times. However, the chances of realistically profiting from individual mining are diminishing.

For example following URL; https://www.nicehash.com/profitability-calculator

The blockchain mining process is configured so that if more miners are digging, the difficulty level goes up, while a decline in the number of miners eases the difficulty level. The higher the mining power in the entire network, the more difficult it becomes for miners to mine a new block because the Bitcoin program adjusts the mining difficulty.

As mining pools represent more significant groups of miners, they play an essential role also a potentially dangerous position in the network. To guarantee the security and trustworthiness of the Bitcoin network in the long term, none of the mining pools (or the combination of a few pools) should dominate the market. When one or more collaborating pools gain the majority of the total mining power, they can perform a 51% majority attack and decide to validate invalid transactions. Only a few large and persistent mining pools can survive as significant and specific investments are needed to mine successfully. The danger of a 51% attack is real. The potential risk raises questions on which factors lead to mining pool concentration and how miners collectively react when concerned about an attack.

[4] The phenomenon of crypto-assets is digitally representing anything that we regard as anything having value putting encryption around it and putting it in the distributed ledger where the ownership of that asset can be assigned and transfer ownership can take place in viable ways.

8.2 Theoretical Background

8.2.1 What Is a "Bullshit Job"?

The word is quoted from David Graeber. His definition for

> a bullshit job is a form of paid employment that is so completely pointless, unnecessary, or pernicious that even the employee cannot justify its existence even though, as part of the conditions of employment, the employee feels obliged to pretend that this is not the case. (David, 2018)

If such a job is a typical bullshit job, why do we kind of mine in front of a computer which could increase productivity? C. Marx believes that capitalism needs a bullshit job to keep it running. However, if such a bullshit job of mining could change this system that keeps us working, it would be necessary to spend our time destroying this system.

Labor benefits are for obtaining a livelihood (salary). Converting labor into capital is detrimental to recognizing the work one really wants to do, such as comfortable jobs. It is not a scientific perspective to discuss whether you like your job. We do not go into the psychological point of view here. Blockchain mining is verified by the node every 10 min. Here we need to identify between the volatile price trends of crypto-assets and the various pioneering management cases of inspired blockchain technology. The point is that we need to investigate the management cases for their suitability, not instability. Zheng, Xie, and Dai roughly categorize the blockchain applications into five domains, which are 1. finance, 2. IoT, 3. public and social services, 4. reputation system, and 5. security and privacy, see Fig. 8.1 (Zheng et al., 2018). Therefore, it is necessary to reconfirm the significance of data

	Fields	brief description
Blockchain Applications	**1. finance**	• Financial Services • **Enterprise Transformation** • P2P Financial Market • Risk Management
	2. IoT	• E-business • Safety and Privacy
	3. public and social services	• Land Registration • Energy Saving • Education • Free-Speech Right
	4. reputation system	• **Academics** [10] • **Web Community**
	5. security and privacy	• Security Enhancement • Privacy Protection

Fig. 8.1 Representative application domains of blockchain (Created by the author)

confirmation in blockchain other than letting the computer data encryption that this work is not Bullshit.

Gordon pointed out that the lives and productivity of most Americans had changed to an enlargement that would have been previously unimaginable between 1870 and 1940. The impact of mechanization, urbanization, automation, etc. shifted many into jobs that paid better but gave them on demand, increasing the amount of time they had to pursue activities other than working, enabling them to receive consumer goods (Gordon, 2017). What is essential is that productivity was promoted by the technology developed at that time and the extra time gained was used for consumption activities. So is consumption the job we are looking for?

8.2.1.1 DAO (Decentralized Autonomous Organization) as a Reputation System

Blockchain technology needs to be reliable for those who want to use it. Therefore, the reputation system in the blockchain application will be described, which is the abovementioned fourth factor. Google has built a system to search for useful information on the Internet with the well-known PageRank. Other social network services (SNS) analyze the relationships within this network and use them to provide the information that users want.

Mitari examines the relationship between the fame of works of art such as paintings and music with the regions (UK, Italy, Germany, and the United States) by age group. Again, this study reveals that fame is associated with social networks.

- What is important here is that the shape of the network ensures the suitability, not instability of the blockchain system. This is because a system structure called DAO (Decentralized Autonomous Organization) constructs its reputation and consensus, which are led by the governance (control).

8.2.2 Why Is Weber's Bureaucracy So Appealing?

For example, suppose a country's military system is inefficient. The military system is a typical bureaucracy. Therefore, the flow of sub-, sub- sub-, and sub- sub- subcontracting jobs to other private companies aims to reform the inefficient bureaucracy. It's considered to be exactly the privatization of the bureaucratic system. M. Weber uses the concept of legitimacy to ensure that the law is essential in modern society. This legitimacy includes the concept of democracy. From this point of view, Freeman's stakeholder theory can be adopted for legitimacy (Freeman et al., 2001). Stakeholder theory has been developed to counter this dominant mindset, such as the bureaucratic system. One side considers the centralized authority in Weber's bureaucracy as a client/server model, and the other considers Freeman's stakeholder theory as a network comparison with a decentralized P2P model. The decentralized P2P model has control methods such as public, private, and consortium blockchain (Jaime & Alfonso, 2011).

The "centralization of information" by huge global ICT companies such as GAFA has reached various technological limits. To do this, we should return to the original Internet with decentralization by DLT technology (George, 2018). The Internet service, which is created by decentralized network architecture, is free, but there is such a crisis here if you want to conduct commercial use, such as financial transactions. Therefore, in order to use the Internet for commercial use, Internet security is essential to protect privacy data. This requires enormous costs to protect the "walled garden." DLT brings the solution for this situation. In other words, the information is fixed from the beginning and created as a database to not be changed.

Recently, several blockchain projects have been created to transform the government system in place in more than 30 countries. This government blockchain is a technology that is directly related to the social system approach. The consensus mechanism forms the core of blockchain technology (Yermack, 2017). Traditionally, a consensus mechanism is not in the field of machinery, but in humankind. Blockchain operates through various types of consensus algorithms with human intervention (MyungSan, 2018). He defines the nature of bureaucracy as a social technology that works as an "information processing machine" for the community. In other words, bureaucracy is a social technology dedicated to distributing and processing information needed in a specific community. It could create a consensus by the multi-stakeholder system (Shimizu, 2016). Therefore, it is theoretically not problematic to claim that the blockchain technology would replace the role of bureaucracy.

8.2.3 Did IoT (Industrie 4.0) Make Us Happy?

Keynes, J. M. said that predicted in the 1930s after World War I, the standard of living of progressive countries 100 years later (1930s + 100 years = 2030) would be four to eight times higher than that time. The world's population has more than tripled from 2 billion in the 1930s to 7 billion in the 2020s. Indeed, his conjecture is correct, given that food and energy allocations are optimized (Maynard, 1930).

Organizations such as the bureaucracy mentioned above have improved our standard of living. Given this, David G. "Bullshit Job" is, What does it mean? Is capitalism just keeping humans working? Table 8.1 The technological developments and working style below describes technological developments that have improved living standards.

Is Blockchain Mining to ensure data trust (such as Privacy data protection, GDPR) described below as a way to automate "Bullshit Job"? Before humans become cyborgs via AI, there is an idea from IoT (Internet of Things) to IoA (Internet of Ability) (Norbert, 1950). Applying this idea, IoA automates the "Bullshit Job."

Table 8.1 The technological developments and working style

Year	brief description
1700	**Manual Labor production** flexibility was highest when manual labor dominated production. "High-mix low-volume" production.
1800	**The first industrial revolution (mechanization)** was marked by a transition from manual labor to the water- and steam-powered mechanical loom. Steam engines were powered by coal and coal mining.
1900	**The second industrial revolution** followed the introduction of electrically powered mass production based on the division of labor. It increased unemployment, as machines replaced many factory workers. The factory acts were passed by the parliament in U.K.
1960	**The third industrial revolution** also known as the digital revolution, occurred after the two great wars. High-speed computers were required during the cold war. The most important innovation was the internet. It was created by the pentagon under the name ARPANET.
1970	**The concept of (AI)** was emerged. It is an entirely different period when computer technology used as an interface. The importance of human power decreased, and production speed increased considerably through computer systems.
1980	**Machine learning** is a subset of the broader field of AI that focuses on teaching computers how to learn without needing to be programmed. To educate a machine, three components are required: Big data-set, high-speed computer and a first algorithm.
2010	**Deep learning** is an advanced class of machine learning. Different types of algorithm exist inspired by the human brain's neural networks. Since 2015, the central processing unit (CPU) has had a counterpart in the graphics processing unit (GPU). Deep learning adopted the structure of the human neural network, using CPU and GPU.
2022 third quarter	**Blockscale**; Intel launches new bitcoin mining chip. Bitmain's Antminer S19 pro has a hash rate of 110 TH/s and consumes 3250 watts of power, and Blockscale chips would have a total hash rate of 148.5 TH/s and consume the same energy efficiency.[a]

Kazuyuki (2020) (Created by the author)

[a]Global Bitcoin Mining Data Review—Q1 2022, URL; https://bitcoinminingcouncil.com/wp-content/uploads/2022/04/2022.04.25-Q1_2022_BMC_Presentation.pdf

8.3 Mining Work

8.3.1 The Geographic Shift of Bitcoin Mining

We have already introduced Satoshi's mining activity by the introduction. What about commercial miners doing large-scale mining? 1660 firms operating in multiple jurisdictions reported their activities in 2020. According to the Cambridge Alternative Finance Center (CCAF), China was the preferred mining site until 2020, but due to a strong regulatory ban in the first half of 2021, most of the operations were suspended, and in 2022 the United States and Canadian mining companies are increasing their presence (Cambridge Center for Alternative Finance, 2021a, 2021b). One of the reasons for China's mining ban may be the problem of using abundant coal to generate energy. Similarly, a giant ICT company such as GAFA uses a lot of electricity to manage their huge data center to utilize and develop artificial intelligence.

Bitcoin reached one trillion US dollars market cap in 2022. One of the largest bitcoin mines is Whinstone US in Rockland, Texas, in America. Whinstone US has measured its 300 MWh[5] per year in developed capacity. Their subsidiary Riot currently has a deployed hash rate capacity of 3.0 EH/s[6] utilizing approximately 91 MWh of energy. Global data center electricity use in 2020 was 200–250 TWh or around 1% of global final electricity demand. This excludes energy used for crypto-assets mining, which was around 100 TWh in 2020 (The IEA, 2022).

8.3.2 Iceland as a Mining Site

Küfeoğlu, S. and Özkuran, M. suggest the historical peak of power consumption of bitcoin mining during the several-week period of time on 18 December 2017 with a demand of between 1.3 and 14.8 GW. This maximum demand figure was between the installed capacities of Finland (~16 G.W.) and Denmark (~14 G.W.) (Küfeoğlu & Özkuran, 2019). With a world population of about 7 billion, electricity demand in developed countries is much higher than in emerging countries. Based on Wikipedia calculations, it is characteristic that the population of about 5.5 million in Finland (relative to the world population, 0.7%) is the fifth largest in the world in terms of average power per capita (1740 watts per person). In particular, Iceland (5898 watts per person)[7] has the highest average power consumption per person, and Norway (2648 watts per person) is the second (Wikipedia). Bitcoin mining consumes a lot of power, and a place with low temperature is preferred for the computer to work properly (Nakamoto, 2018).

Science said data centers are energy-intensive enterprises, estimated to account for around 1% of worldwide electricity use. These trends have clear implications for global energy demand and must be analyzed rigorously. The worldwide energy use of data centers had grown from 153 terawatt-hours (TWh) in 2005 to between 203 and 273 TWh by 2010, totaling 1.1–1.5% of global electricity use. Since 2010, however, the data center landscape has changed dramatically. By 2018, global data

---

[5] Megawatts per hour are used to measure the output of a power plant or the amount of electricity required by an entire city. For example, a typical coal plant is about 600 MWh, which means Whinstone U.S. uses half of a coal plant. The unit of electricity is watts per hour, kilowatt per hour, megawatts per hour, gigawatt per hour and terawatt per hour. For example, New, energy-efficient refrigerators use about 300–400 kilowatt-hours per year.

[6] Hashrate is the speed of mining. A unit of hash/second means how many calculations per second can be performed. Machines with a high hash power are highly efficient and can process a lot of data in a single second. Hashrate is usually measured in units of k (kilo, 1000), M (mega, one million), G (Giga, one billion), T (tera, 1 trillion), P (Peta, one quadrillion), or E (Exa, one quintillion).

[7] According to the government of Iceland, about 85% of the total energy supply in Iceland is derived from domestically produced renewable energy sources. This figure is the highest share of renewable energy in any nation.

center workloads had increased more than sixfold, whereas data center Internet protocol (I.P.) traffic had increased by more than ten-fold. But since 2010, electricity use per computation of a typical volume server has dropped by four, mainly due to processor-efficiency improvements. These data suggest that, although global data center energy consumption has increased slightly since 2010, energy growth has been substantially decoupled from growth in data center compute instances over the same period (Masanet et al., 2020). This refers purely to data center usage for cloud data services and reflects the situation prior to the introduction of ChatGPT (November, 2022).

8.3.3 Which Is More Valuable, Data Or Coal Mining?

According to the CCAF mentioned above, China was the most significant mining site until 2020, but most operations were suspended due to a strong regulations ban in the first half of 2021. And many Chinese miners moved to the United States and Canada in 2022. Akyildirima E., Corbet S. and Luceye B. suggest that global coal prices, especially Chinese coal mining, are more linked to Bitcoin price fluctuations than real mining conditions (Akyildirim et al., 2021). The reason for the results of this paper is the use of the Dynamic Conditional Correlation (DCC) model, which considers systematic risk as the basis for its calculation. The DCC model systematically measures other effects. If environmental problems are ignored, the price of coal as a commodity is subject to price fluctuation factors due to Blockchain Mining. In other words, if this frees us from "Bullshit Jobs," rational market participants rely more on value-added data mining than on natural resource (coal) mining. In addition, the basic Bitcoin ownership structure is that 6.28% of addresses own 93.72% of total assets, which is a distribution like Pareto optimal (Dániel et al., 2014). Bitcoin's anonymity and visualization of transaction history are different from banking transactions that involve centralized control as an institution.

However, the ownership structure mentioned above increases the utility of large ownership. For mining and other management applications, a new way of incentive methods such as reputation applies to increase the utility of decentralization (Shimizu, 2019).

8.4 Essential Mining

Even if the giant ICT companies such as GAFA's dominance changes somewhat with blockchain technology, it is unclear whether our way of the bullshit job will change. However, the mental struggle to escape and win freedom from such a dominant situation seems healthy.

> Markov chains tell you the statically likely future without knowing the past; blockchains enable the real-life future by indelibly recording the past. (George, 2018)

The future in which distributed ledgers support decisions is more independent than AI (artificial intelligence), which disconnects probabilistic states. The difference between a bullshit job and an essential job is trust. For example, data mining is a term similar to digging gold, because currency as a payment method requires the trust of "gold." This trust recognizes currency = Bitcoin. Where is the trustworthiness of digital data in this sense? Once you trust a trustworthy database, even if data mining is bullshit mining, the data that supports decisions can be trusted, and mental stability can be rewarded.

8.4.1 IoT to IoA (Internet of Ability)

Usually, the Internet means a personal computer identifies and exchanges digital information in the network. This identification is mainly made by people who use these PCs. In this sense, IoT (Internet of Things) means a thing positions in the network. IoT began to be used in a study called RFID (Radio Frequency Identification) at MIT for the first time in 1999. Identification-product management, which uses barcodes was positioned on the Internet using these radio waves. Also, IoT has a very similar meaning to ubiquitous (everywhere) computing (see in Fig. 8.2).

Blockchain technology is mainly utilized for data management operations in healthcare and IoT, specifically to improve data security, including data integrity, access control, and privacy protection. Blockchain and IoT, including health IoT, can have similar meanings on the IoA (Internet of Ability). The most recognized example of an IoB is a pacemaker, a small device placed in the chest to help patients with heart conditions control abnormal heart rhythms with electrical impulses (Adere, 2022).

	Fields	*Example*
IoT (Internet of Things)	Supply chain	RFID, Connected Cars
	Automate "Bullshit Jobs"	
	Human Augumentation	HappinessCounter (^^)
IoB (Internet of Boby)	Healthcare, Medical equipment	Wearable health monitors
IoA (Internet of Ability)	Business collaboration, innovation	Supply chain

Fig. 8.2 Automation of "bullshit job" through the human augmentation (Created by the author)

8.5 Conclusion

C. Frey & M. Osborne investigated jobs that are replaced by the supercomputers such as those using artificial intelligence. They estimate that about 47% of total US employment is at risk (Osborne & Frey, 2013). Their suggestion shocked many workers. Should we seriously feel uncomfortable being "meaningful job" replaced by AI? Aren't there more "Bullshit Jobs" out there than you have to do?

1. First, clarify the "Bullshit job" in the bureaucracy,
2. Clarifying the consensus system by DAO.
3. The important thing here is to accept the instability of the blockchain system and find compatibility.
4. Through that, create an incentive system called reputation, which people believes.
5. Reputation isn't a way of "large cargo" incentive, that is a value transformation.
6. Instead of probabilistically creating a reputation from conventional analogical data, create it with data trusted by blockchain.
7. This helps DAOs (Decentralized Autonomous Organizations) use reputation to find governance structures.

8.6 Review Questions

1. What is a "bullshit job," and in what way can blockchain mining be seen as one? Can it also create new social or economic value?
2. How does a DAO (Decentralized Autonomous Organization) build trust through a reputation system, and how is this different from traditional bureaucratic legitimacy?
3. How has technology—from manual labor to IoT and IoA—changed human work from physical tasks to decision-support roles?
4. Why have mining sites moved from China to the United States and Iceland? What does this show about global governance and regional technology levels, such as semiconductor or energy efficiency?
5. Markov chains predict the future without remembering the past, while blockchain records the past to shape the future. How might this difference affect future governance and decision-making systems?

References

Adere, E. M. (2022, July 14). Blockchain in healthcare and IoT: A systematic literature review. *Array*. https://doi.org/10.1016/j.array.2022.100139

Akyildirim, E., Corbet, S., & Lucey, B. M. (2021). China, coal, calamities and cryptos. *SSRN*. Retrieved from https://ssrn.com/abstract=3851253

Cambridge Center for Alternative Finance. (2021a). *The 2nd global alternative finance market benchmarking report*. Retrieved from https://www.jbs.cam.ac.uk/faculty-research/centres/alternative-finance/publications/the-2nd-global-alternative-finance-market-benchmarking-report/

Cambridge Centre for Alternative Finance. (2021b). *Bitcoin network power demand*. Retrieved from https://cbeci.org/

Dániel, K., Márton, P., István, C., & Gábor, V. (2014). Do the rich get richer? An empirical analysis of the bitcoin transaction network. *PLoS One*. https://doi.org/10.1371/journal.pone.0086197

David, G. (2018). *Bullshit Job*. Penguin Random House.

Freeman, E. R., Harrison, J., Wick, A., Parmar, B., & Colle, S. (2001). *A stakeholder approach to strategic management*. Blockwell.

George, G. (2018). *Life after Google: The fall of big data and the rise of the blockchain economy*. Regnery Gateway. Retrieved from https://www.amazon.co.jp/Life-After-Google-Blockchain-Economy/dp/1621575764

Gordon, R. J. (2017). *The rise and fall of American growth: The U.S. standard of living since the civil war*. Princeton Univ Pr.

Human Augmentation Research Initiative. (2019). *Human augmentation research*. Retrieved from https://humanaugmentation.jp/

Jaime, G.-J., & Alfonso, G.-C. (2011). Overview and challenges of overlay networks: A survey. *International Journal of Computer Science & Engineering Survey*. Retrieved from https://www.researchgate.net/publication/50199321_Overview_and_Challenges_of_Overlay_Networks_A_Survey

Shimizu, K. (2020, October). Digital transformation of work and ESG: Perspectives on monopoly and fair trade. *Risk Governance and Control: Financial Markets & Institutions*, 75–82. https://doi.org/10.22495/rgcv10i3p6

Küfeoğlu, S., & Özkuran, M. (2019, May). *Energy consumption of bitcoin mining*. https://doi.org/10.17863/CAM.41230

Masanet, E., Shehabi, A., Lei, N., Smith, S., & Koomey, J. (2020, February). Recalibrating global data center energy-use estimates. *Science*. https://doi.org/10.1126/science.aba3758

Maynard, K. J. (1930). *Economic possibilities for our grandchildren*. Retrieved from http://www.gutenberg.ca/ebooks/keynes-essaysinpersuasion/keynes-essaysinpersuasion-00-h.html#Economic_Possibilities

MyungSan, J. (2018). Blockchain government—A next form of infrastructure for the twenty-first century. *Journal of Open Innovation: Technology, Market, and Complexity*. https://doi.org/10.1186/s40852-018-0086-3

Nakamoto, S. (2018). *Bitcoin: A peer-to-peer electronic cash system*. Bitcoin.org. Retrieved from https://bitcoin.org/ja/bitcoin-paper

Norbert, W. (1950). *The human use of human beings; Cybernetics and society*. Retrieved from http://21stcenturywiener.org/wp-content/uploads/2013/11/The-Human-Use-of-Human-Beings-by-N.-Wiener.pdf/2014.05.10

Osborne, M., & Frey, C. B. (2013, September). *The future of employment: How susceptible are jobs to computerisation*? Retrieved from https://www.oxfordmartin.ox.ac.uk/downloads/academic/The_Future_of_Employment.pdf

Shimizu, K. (2016). Socio-cybernetic approach into the triumvirate: Stakeholder governance between management, shareholders and employees. *An Enterprise Odyssey. International Conference Proceedings*, 273–280. Retrieved from https://search.proquest.com/docview/1815362010?pq-origsite=gscholar&fromopenview=true

Shimizu, K. (2019). "Pricing game" for tacit collusion and passive investment. . Retrieved from http://ceur-ws.org/Vol-2348/paper23.pdf.

The IEA. (2022). *Data center and data transmission networks*. Retrieved from https://www.iea.org/reports/data-centres-and-data-transmission-networks

Wikipedia. (2022). *List of countries by electricity consumption*. Retrieved from https://en.wikipedia.org/wiki/List_of_countries_by_electricity_consumption

Yermack, D. (2017). Corporate governance and blockchains. *Review of Finance*, 21(1). https://doi.org/10.1093/rof/rfw074

Zheng, Z., Xie, S., Dai, H.-N., Chen, X., & Wang, H. (2018, October). Blockchain challenges and opportunities: A survey. *Article in International Journal of Web and Grid Services*. https://doi.org/10.1504/IJWGS.2018.10016848

Chapter 9
The Countervailing Power of AI DAOs Influences Value Transformation: Bitcoin (PoW) vs. Ethereum (PoS)

Abstract In this paper, we investigate the process of value transformation influenced by the countervailing power of decentralized autonomous organizations (DAO) controlled by artificial intelligence (AI). There are various values in society. The Internet was believed to lead humanity in a better direction by further decentralizing various values. However, the uneven distribution of information causes many problems, such as "cyber cascades," "filter bubbles," and "echo chambers," etc. To solve these problems, DAO using blockchain is expected to become a method for solving these problems in the Internet space.

This paper intends to capture the value change in three steps. In the first step, we take two philosophical approaches. First, using Hegel's dialectic, we attempt to compare A. current social values (thesis), B. the value of, for example, AI DAOs (antithesis), and C. new values (synthesis) in an Internet world where these contradictions exist (A + B = C). The second philosophical approach is to consider the mutuality of Bitcoin (PoW) and Ethereum (PoS) by adapting Popper's World 1, 2, and 3 models.

In the second step, we examine the three core capabilities that amplify "credit" in decentralized finance (DeFi): exchange swap rates, staking, and indices (portfolio). Finally, we consider the value dispersion within the Bitcoin and Ethereum networks as the main two poles. The countervailing power between these two poles and the upcoming expanding AI DAOs, especially Bitcoin-halving, coordinate the interests of stakeholders due to the uneven distribution of values and encourage the interaction between diverse values in cyberspace.

Keywords The countervailing power · AI · DAO · Web3 · DeFi · artificial society

K. Shimizu, *ESG Investment, German Industrie 4.0, and Blockchain*,
https://doi.org/10.1007/978-981-95-6927-4_9

9.1 Methodology

Humanity has various values: legitimacy, human rights, freedom, security, dignity, and human life. First, consider two methodologies. Dialectics considers matter to be the root of spirit. Hegel's dialectics: For example, consider value relationships using this dialectic. Also, we applied Popper's World 1/2/3 theory to the current value relationships. Idealism emphasizes the compositional ability inherent in the mind rather than matter.

- Based on **Hegel's dialectics**, A. Thesis: ex. "openness of blockchain" + B. Antithesis: "private information" = C. Synthesis: "using hybrid technology of public and private chain."
- **Popper's Worlds I, II. III**; Interaction between World I: the "openness of blockchain"~ the dilemma between World I and World II: "private information," + World III: the new Web 3 interaction with the captured values. (Karl, 1978).

The global crypto market cap is $1.16 T in June 2023. As mentioned above, Bitcoin's dominance is around half of the total crypto market (46.53% in June 2023). The share of Ethereum is around 19% (Omkar, 2023). The total crypto market trading volume is $32.26B. The total volume in DeFi is about 8% ($2.37B), and the volume of all stablecoins is about 92% ($29.69B) of the total crypto market. We will examine the concept here with PoW, PoS and the rest of the consensus algorithms for simplicity and understanding.

- Based on Hegel's dialectics, A. Bitcoin's PoW (thesis) + B. Ethereum's PoS (antithesis) = C. Co-evolution of both consensus algorithms (synthesis). The simple formula is A + B = C.
- Popper's Falsifiability and Worlds I, II. III; Bitcoin's PoW (world 1), Ethereum's PoS (world 2), and co-evolution of both consensus algorithms (world 3). Bitcoin's PoW (world 1), Ethereum's PoS (world 2), and co-evolution of both consensus algorithms, such as NPoS (Nominated Proof of Stake), PoH (Proof of History), PoA (Proof of Authority), PoC (Proof of Consensus), etc. (world 3).

Diagram of knowledge growth by Popper's idea:

P_1 (Problem 1)—TT (Tentative Theory)—EE (Error Elimination)—P (Problem 2); See below Fig. 9.1 The logic tree of this paper.

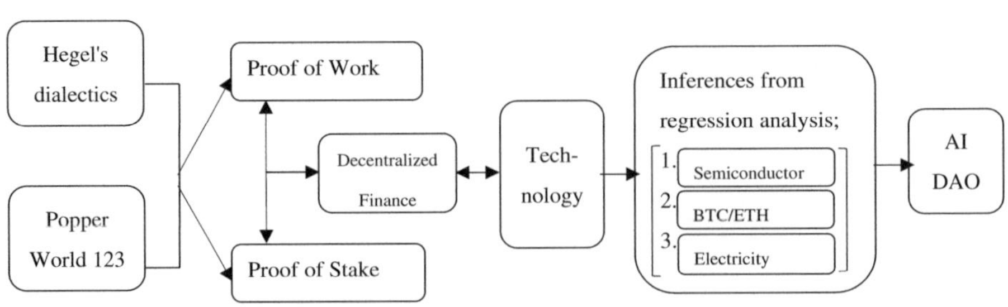

Fig. 9.1 The logic tree of this paper. Source: by the author

9.2 DeFi (Decentralized Finance) as a Value Accelerator

9.2.1 Brief History of DeFi

What we look for in a financial institution is trust. For this reason, financial institutions are also called credit institutions. Financial institutions are trusted because they can settle transactions smoothly. The source of trust in financial institutions is that they do not lie. A similar trust is formed if this source of trust is unbreakable cryptography. The PoW blockchain project, Bitcoin, secures this trust.

Personal overseas money transfers and on-demand transactions became possible, enabling several online value transformations. As a result, cryptocurrencies have significantly impacted traditional centralized financial (CeFi). However, decentralized finance (DeFi) in the early 2000s was still in the prototype technologically, and its business application had just begun (Nakamoto, 2008). Ethereum was developed by Vitalik Buterin in 2013 as the research and development of blockchain technology progressed (Buterin, 2014). By incorporating smart contracts based on cryptocurrencies, blockchain technology can be used for various purposes.

Furthermore, by changing Bitcoin's Proof of Work (PoW) problem to Proof of Stake (PoS), the problem of energy consumption can now be addressed. By decentralizing such diverse values, the Internet was believed to lead society in a more pluralistic and better direction. To solve such issues, DAO on Web3, powered by blockchain, is expected to promote the democratization of the "human" internet space (Simon, 2023) (Nielsen, 2022).

9.2.2 Orderbook vs. Liquidity Pool at Leyer 1 (See in Table 9.1)

Uniswap is a decentralized exchange (DEX) and part of the decentralized finance (DeFi) product ecosystem launched on Ethereum mainnet in November 2018. It replaces the traditional order book type trading on centralized exchanges (CEX). At a very high level, an AMM (Automated Market Maker) replaces the buy and sell orders in an order book market with a liquidity pool of two assets valued relative to each other. As one asset is traded for the other, the relative prices of the two assets shift and a new market rate for both is determined. In this dynamic, a buyer or seller trades directly with the pool rather than with specific orders left by other parties (Uniswap, 2023). Liquidity providers can be regarded as investors in the decentralized exchange and earn fixed commissions per trade. They lock up funds in liquidity pools for distinct pairs of currencies allowing market participants to swap them using the improved price function. Liquidity providers take on market risk as liquidity providers in exchange for earning commissions on each trade. In short, Investors as a customer accept market risks usually taken by traders. This new pool trading has a risk profile of a liquidity provider and the so-called impermanent (unrealized) loss in particular (Andreas & Gurvinder, 2021).

Table 9.1 A pattern of AMM

AMM	Connection of individual tokens
Uniswap	X * Y = K
Balancer	(X * Y * Z) ^ (1/3) = K
Curve	(X * Y) + (X + Y) = K

Source: JBA DeFi study group (Created by the author)

Table 9.1 below explains the following; **Uniswap** introduced the constant product market maker formula to ensure continuous liquidity in exchanging tokens on Ethereum. The formula follows: x is token 1, y is token 2, and k is a constant.

Curve's primary distinction from other decentralized exchanges, such as Uniswap, is its low slippage and low fee algorithm specifically designed for trading between assets of the same value. Curve makes it very useful for stablecoin swaps, as they are expected to hold roughly the same value (Perpetual Protocol, 2022).

Balancer's pools are like index funds that construct a portfolio of assets with fixed weights. The balancer protocol allows each pool to have two or more assets and to supply them in any ratio. Each asset reserve is given a weight when the pool is created and the weights sum to one.

9.3 Bitcoin (PoW) vs. Ethereum (PoS)

The competitive principle of getting Coinbase through mining functions as a solution to the 51% problem (double-spending problem). After mining, the BTC transaction ledger is recorded and held as a single block on c.a. 16,000 nodes worldwide, and these nodes confirm this block once every 10 min. This is the final approval of transaction information on the final distributed ledger. This approval method is based on a block number (nonce: a hash value with approximately 16 leading zeros) that is encrypted using a consensus-building process (perfectly competitive authentication process) by an unspecified number of participants called PoW (Proof of Work). It is a competition for discovery. This discovery competition is held worldwide, and because it is a winner-take-all system, the problem of wasted energy use has been pointed out. Considering the community debate surrounding the future trajectory of Bitcoin versus Ethereum consensus protocols, there's ongoing discussion about the feasibility of transitioning Bitcoin to a Proof of Stake (PoS) mechanism (ethereum.org, 2024). While Ethereum has successfully implemented PoS, Bitcoin continues to rely on the Proof of Work (PoW) mechanism. (Alex, 2023).

9.3.1 *Consensus Mechanisms (See in Table 9.2)*

Fundamental differences: PoW requires miners to solve complex mathematical problems to validate transactions, while PoS relies on validators who hold a certain amount of cryptocurrency. Here we believe that community consensus applies to the

Table 9.2 Differences between BTC and ETH

Feature	Bitcoin (BTC)	Ethereum (ETH)
Primary purpose	Digital gold, store of value	Platform for smart Contracts and Decentralized applications
Transaction speed	Average 10 min (block generation time)	About 15 s (block generation time)
Transaction cost	Can be high in times of network congestion	Gas fees (variable transaction fees)
Consensus mechanism	Proof of work (PoW)	Transitioning to proof of stake (PoS)
Supply limit	Capped at 21 million BTC	No fixed supply limit (though issuance rate varies)

Source: Alex (2023) and Alistair and Eleftherios (2020), (Created by the author)

idea of the Popper 123 world. The Bitcoin community is divided on whether transitioning to PoS is necessary or beneficial, with some advocating for maintaining the current PoW model.

Concerning energy efficiency, PoS is considered more energy-efficient than PoW, which consumes significant electricity. The gas fee and time (transaction cost) using PoW are much higher than PoS. Security concerns remain. Some argue that PoS may be more vulnerable to attacks or centralization, while others believe it can provide equal or even higher security compared to PoW (Sal, 2019). The consensus is different for both algorithms, which are mentioned above. Ultimately, whether Bitcoin can switch to PoS will depend on the consensus reached within the community and the perceived benefits and drawbacks of such a transition (Stephen, 2015).

“The Merge” of Ethereum on 15 September 2022 fundamentally altered the network’s architecture. The upgrade has considerable implications for the network’s electricity usage, which we estimate to have decreased by a significant 99.99%. This reduction is caused by PoS not requiring the same level of computational power as PoW, as it relies on validators who hold a stake in the network, rather than miners, who utilize powerful hardware to solve cryptographic puzzles.

PoS consensus mechanism is used to ensure trustless, fundamentally different from proof of work. Unlike PoW, PoS substitutes the resource cost associated with the computationally intensive process of guessing random numbers to solve a cryptographic puzzle with a requirement to pledge financial resources in the form of the blockchain’s native tokens as collateral, so-called “staking.” To participate in attesting or proposing new blocks, those who are validators must lock or “stake” a set number of native tokens, for instance, ether (ETH) in the case of Ethereum.

How do people usually give value? We adapt Weber’s concept of authority to consensus algorithms like Bitcoin. Barnard describes the acceptance of authority (the realm of indifference) in relation to standards. This criterion is a “statistically significant difference.” Interests and motivations become a way to unify the boundaries between general and individual values.

9.3.2 Inferences From Regression Analysis

The multiple regression analysis here will explain the dominance situation of BTC and ETH (CoinMarketCap, 2023), using R. Therefore, in this study, the dependent variable is semiconductor sales (WSTS, 2023), and the independent variable is the dominance status of BTC and ETH. Semiconductor sales and the BTC and ETH dominance status (each measured by market capitalization) are based on a 10-year time-series data set from 2013 January to 2023 January, on a weekly basis.

The dependent variable is global semiconductor sales, and the independent variables are the market capitalization of BTC and ETH. As a result, BTC and ETH's market capitalization had 1% significance in predicting global semiconductor sales. Different letters indicate significant differences in the *** test ($P < 0.01$). All data were Z-transformed, but the same results were obtained by inverse Z-transformation.

- Adjusted R-squared: 0.7015, BTC (b = 0.43, SE = 0.07, t(552) = 5.97, P = 4.09e-09 ***),

  ETH (b = 0.42, SE = 0.07, t(552) = 5.72, P = 1.80e-08 *).**

It was confirmed that BTC (PoW) and ETH (PoS) have a 1% significance in terms of semiconductor sales using both consensus algorithms. It seems to confirm the relationship between PoW and PoS consensus algorithms and technological innovations in the semiconductor (silicon) cycle that occur every 4 years.

The dependent variable is Asian semiconductor sales, and the independent variables are the market capitalization of BTC and ETH. As a result, the market capitalization of BTC and ETH significantly predicted semiconductor sales in Asia.

- Adjusted R-squared: 0.6734, BTC (b = 0.68, SE = 0.024, t(552) = 8.78, P = <2e-16 ***),

  ETH (b = 0.16, SE = 0.024, t(552) = 2.09, P = 0.0372 *).

The dependent variable is US semiconductor sales, and the independent variables are the market capitalization of BTC and ETH. As a result, ETH's market capitalization was a strong predictor of US semiconductor sales. However, BTC did not show any significance for US semiconductor sales.

- Adjusted R-squared: 0.6394, **ETH (b = 0.695, SE = 0.08, t(552) = 8.68, P = <2e-16 ***),**

  BTC (b = 0.11, SE = 0.08, t(552) = 1.375, P = 0.17).

We conducted a multiple regression analysis of the relationship between two types of semiconductor sales by region (America and Asia) and BTC and ETH. As a result, the 1% significance in semiconductor sales in the Asian region over BTC was calculated to be higher than the 10% advantage over ETH. Additionally, the 1% significance of ETH in semiconductor sales in the Americas region was calculated to be more significant than BTC.

As pointed out in Dániel Kondor's paper (Dániel et al., 2014), "The Bitcoin system has 6.28% of addresses owning 93.72% of the total assets." BTC mining is done by one specific Chinese mining site, where electricity is cheap. This Chinese consensus is communism and a PoW-like decision-making method, and in the United States it is famous that the richest 1% holds a 90% of total American financial assets.

The WSTS (World Semiconductor Trade Statistics) 2024 Semiconductor Market Outlook, forecast in the fall of 2023, predicts that the global semiconductor market will increase by 13.1% and reach a valuation of US$588 billion. This growth is primarily driven by the memory sector, which is expected to soar to approximately US$130 billion in 2024, an increase of more than 40% year on year. IDC says that from 2024 to 2026, accelerating demand for AI servers and AI-enabled endpoint devices will drive semiconductor content growth and drive new upgrade cycles across the enterprise (Shirer, 2023). Current cloud-based AI processing is inefficient because it requires energy to transfer data because it is processed by a distant system. AI-enabled endpoint devices separate simple and complex information processing and perform most of their processing on general-purpose memory, CPUs, GPUs, and ASICs (application-specific integrated circuits) where data is generated.

9.3.3 Supply and Demand of Semiconductor and Electricity

What does it mean that the Bitcoin system is in a Pareto situation, with 93.72% of the total assets held in 6.28% of addresses? Does BTC mining reflect the geographical consensus of China? BTC's consensus algorithm is PoW. PoW mining involves an unspecified number of participants creating cryptography in perfect competition. This competition of free will makes a system that does not require a third party (Adam, 2007). Prateek Goorha explains gold holding and its circulation from a historical perspective. Building a concentrated inventory of scarce goods is pointless if that inventory cannot make any circulation between social, economic, and political values. He also points out that a rare item has little meaning as a subjective meditated value if the concentration does not change. After all, there are far rarer elements (or technology) on the periodic table than gold (Prateek, 2019). In other words, the act of people exchanging "trust" that establishes accumulated truths has extremely important meaning for economics.

Bitcoin-halving, also known as "Coinbase halving," is a significant event in the Bitcoin network. It refers to the reduction by half of the rewards granted to miners for mining a new block on the blockchain. Initially, miners received 50 BTC per block. This reward halves approximately every 4 years, or after every 210,000 blocks mined. This process will continue until the total supply of Bitcoin reaches 21 million BTC. Halving reduces the rate of new Bitcoin entering circulation, acting as a deflationary mechanism to potentially increase the value of Bitcoin over time due to its reduced supply growth. As of 2024, people are focusing on the aspect of PoW's BTC that the special feature has the issuing cap, being determined in the context of central banks of various countries implementing monetary easing and accelerating global inflation after the COVID-19 pandemic that hit the world (see in Table 9.3).

Table 9.3 BTC halving schedule

Event	Date span	Duration (month)
1st BTC halving	From 03/01/2009 to 28/11/2012	47
2nd BTC halving	From 28/11/2012 to 09/07/2016	40
3rd BTC halving	From 09/07/2016 to 18/05/2020	44
4th BTC halving	From 18/05/2020 to around ??/05/2024	Around 48
5th BTC halving		Latest block no.825,278
	Continue to 2140 every 210,000 blocks	

Source: Hasan (2021) (Created by the author)

BTC mining should not be used in an unsustainable model that ignores environmental issues. Therefore, The Cambridge Bitcoin Electricity Consumption Index (CBECI) analyzes the relationship between BTC and ETH (using PoW) electricity consumption (Cambridge Center for Alternative Finance, 2022). Consequently, we conducted a regression analysis using electricity usage data as a mediating variable on the abovementioned regional semiconductor sales data and market capitalization data of BTC.

We conducted a mediation analysis for the period from September 2019 to January 2022 to see if electricity demand in Asia mediates the impact of trading volume of BTC on semiconductor sales in Asia. First, we conducted a regression analysis using Asian semiconductor sales as the dependent variable and trading volume of BTC as the explanatory variable. As a result, trading volume of BTC had an advantage in predicting semiconductor sales in Asia (b = 0.84, SE = 0.84, $t(125) = 17.39$, $P = 0.00$). Furthermore, by adding power consumption in Asia as an explanatory variable, Asia's electricity consumption significantly predicted semiconductor sales in Asia ($b = -0.51$, SE = −0.51, $t(125) = -7.92$, $P = 0.00$). And the effect of trading volume of BTC also had an advantage (b = 0.44, SE = 0.44, $t(125) = 6.92$, $P = 0.00$). Since this data has been normalized, the standardization coefficient and standard error are the same value.

What is very interesting here is that the coefficient in Asia is negative. This reflects the impact of China's mining regulations since mid-2021 (see Fig. 9.2 DAG (Directed Acyclic Graph) for Asian power supply and demand as mediating effects). If such regulations are excluded, in order to confirm whether the US power demand will lead to the effects of trading volume of BTC, the same period from September 2019 to January 2022 was analyzed. In the US, the trading volume of BTC naturally affects semiconductor sales through electricity consumption. First, a regression analysis was implemented with US semiconductor sales as the dependent variable and trading volume of BTC as explanatory variables. As a result, the trading volume of BTC was significant in predicting US semiconductor sales ($b = 0.73$, SE = 0.73, $t(125) = 11.98$, $p = 0.00$). Furthermore, if the US power consumption (AmeEle) was added as an explanatory variable, the US power consumption was greatly predicted

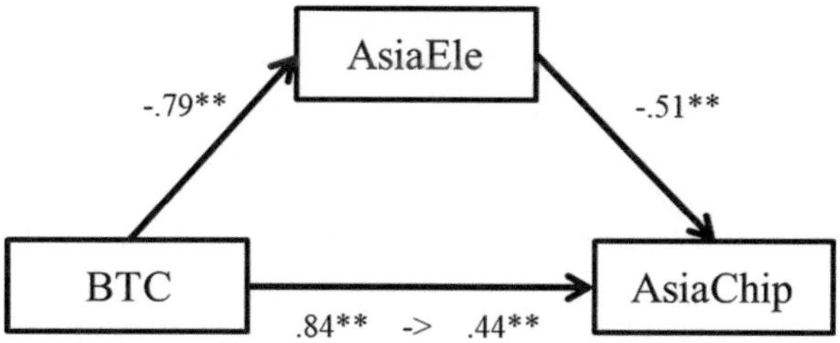

Fig. 9.2 DAG (Directed Acyclic Graph) for Asian power supply and demand as mediating effects. Source: HAD (Created by the author)

US semiconductor sales, AmeChips (b = 0.64, SE = 0.64, t(125) = 9.2, p = 0.00.). The effect of trading volume of BTC was also significant advantageous (B = 0.52, SE = 0.52, t(125) = 8.18, p = 0.00). Since this data were normalized, the standardization coefficient and the standard error are the same value.

The case in the United States reflects the trade relationship between the United States and China, which was also pointed out in The Cambridge Bitcoin Electricity Consumption Index (CBECI). The CBECI indicators indicate the relationship between semiconductors and power used as hardware. It is ETH's merge that changed this relationship.

9.4 AI DAO, Artificial Societies and Post-truth

Following Hegel's dialectics, we consider the current values (thesis), the AI DAOs values (antithesis), and the new values (synthesis) in which the two are in a countervailing relationship in the Internet world (Stanford, 2020). In this research, we seek a procedure for sustainability that transcends human interests and seeks technocracy ("rule by mechanism") advocated by Thorstein Veblen, John K. Galbraith, and others (Galbraith, 1952). Specifically, governance by AI and DAO. The issues of privacy, monopoly/oligopoly, human rights, freedom, security, dignity, and human life thus require changes in our value orientation (Yamazaki et al., 2020). There are two value distribution models: the centralized client–server and decentralized P2P models. In the former, AI is at the center and distributes (unevenly) value (Freeman et al., 2001). On the other hand, in the P2P model, DAO operated by various AI agents and values by humans are distributed. Distributed values create real and artificial societies that are interpreted by AI agents created through blockchain programs in cyberspace.

In natural ecosystems, many animal populations are self-organized. For example, the behavior of a single ant is simple and limited, but ant colonies automatically control complex behaviors such as foraging, feeding, nest building, and defense. Significantly, individuals forming ant colonies can adjust according to the division of labor and their behavior based on environmental changes. In addition to ants, fish, bees, and swarms also exhibit similar self-organizing behaviors. All these animal populations are characterized by centralized control and a lack of hierarchy. In other words, it shows that group behavior on the Internet can be controlled autonomously (Shuai et al., 2019).

Also, in this network, traditional values, AI programs, and his DAO democratic compete and interrelate (Computer Politics; Algocracy). This is governance based on mutual relationships (value chains) with DAOs and AIs and countervailing power (Jacques, 1999). Schumpeter believes that innovation is fostered by entrepreneurship. Furthermore, considering economic cycles, we can assume a 4- to 50-year cycle of Kitchin, Juglar, and Kondratiev (Joseph, 1939). The short-term Kichin cycle is consistent with the silicon cycle. While Hegel's materialist view of history and Marx's historical materialist view of history have an influence on people's way of thinking, the ideas of idealism, such as Popper's, have also had an influence.

Figure 9.3 above shows the relationship between semiconductor sales and BTC market capitalization in chronological order. Then, when explaining the explained variable (SemiChips) using the explanatory variable BTC, the multiple regression coefficient is 0.69, which has moderate explanatory power, and the regression equation is $Y = 0 + 0.38x$, and BTC and SemiChips are 2e-16 ***, which are highly significant.

The problem of blockchain's openness and "privacy" can influence the recognition of the technology itself; anonymity destroys all possibilities for Post-truth understanding (Kiyoshi & Yohko, 2021). However, value transformations are occurring through Web3 via Blockchain in privacy, monopoly/oligopoly, human rights, freedom, security, dignity, and human life. Our collective intelligence and the countervailing power of AI DAOs using blockchain technology will form this value transformation (Humans.ai, 2022). Its value transformation process is carried out in three stages according to the following Hegelian dialectics and Popper's world 1/2/3 model;

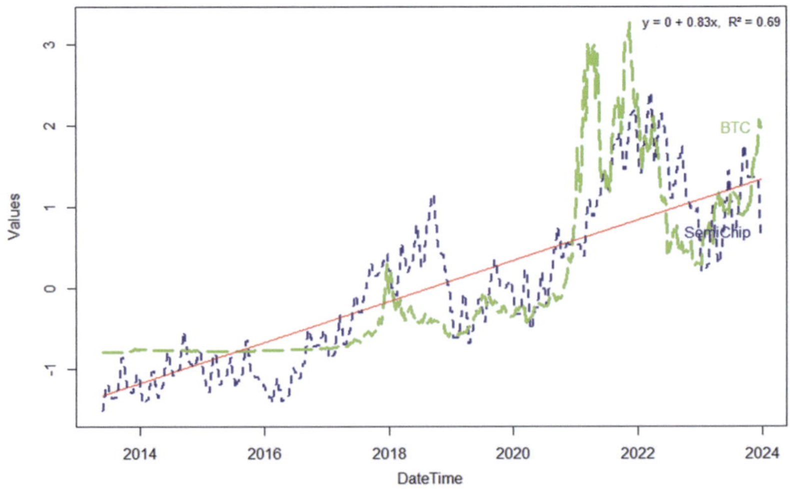

Fig. 9.3 SemiChips and BTC correlation and regression graph. Source: R (Public, n.d.) (Created by the author)

1. The current values (thesis), such as the dilemma with "privacy" and PoW vs. PoS (Shimizu, 2021).
2. The AI DAOs' values (antithesis) and Post-truth, such as appropriate information, are provided to professionals and lead to appropriate decisions.
3. The new values (synthesis), in which the two are in a countervailing power in the Internet world. In this network, traditional values, AI programs, and his DAO democratic compete and interrelate (Computer Politics; Algocracy).

The most important characteristic of blockchain is "creditability" due to its "openness," which creates a dilemma with "privacy." Therefore, the value transformation in the "privacy"issue in blockchain varies according to the concepts of Hegel and Popper. In general, privacy is divided into a wide range of contents. Two of the most important issues are "credit" and "falsification." This paper focuses on the characteristics (limitations) of blockchains that expose privacy. The countervailing power of values and the elimination of value boundaries coincide with the fading of the border between real- and cyberspace. The logic is that Web3 solves the problems of the traditional concept of privacy and also creates a new way of thinking (Post-truth). The "openness and trust" of blockchain prevent "lack of trust" and contribute to better decisions (Bernd et al., 2022). The two ideas of Post-truth interact, and it would be more natural to think of them as an interaction rather than as polarities between PoW and PoS. As a future research topic, we will continue to monitor the progress of PoW and PoS with great interest. If AI-like capabilities were to amplify value automatically, we would have to encourage economic circulation to take place and be used in human interaction.

9.5 Review Questions

1. What does the countervailing power of AI DAOs mean, and how can it transform social and economic values?
2. How does decentralization in blockchain differ from the centralized information systems used by companies like GAFA?
3. How can DAOs help solve Internet problems such as filter bubbles and echo chambers?
4. From the perspective of Hegel's dialectics (thesis–antithesis–synthesis), how can the tension between Bitcoin (PoW) and Ethereum (PoS) create new forms of value?
5. How does Popper's World 1–2–3 theory explain the material, mental, and cultural development seen in blockchain technology?
6. How does trust differ between centralized finance (CeFi) and decentralized finance (DeFi) systems?
7. Why was the Ethereum Merge (2022) considered an important event for reducing the energy consumption of blockchain networks?
8. How are the market capitalizations of Bitcoin and Ethereum statistically related to global semiconductor sales?

9. What does the positive correlation between BTC/ETH and semiconductor demand suggest about ongoing technological cycles such as silicon or AI development?
10. In what ways is AI DAO governance similar to the self-organizing behavior found in nature, such as in ant colonies or bee swarms?

References

Adam, S. (2007). *An inquiry into the nature and causes of the wealth of nations* (Vol. 1, 2, 3, 4 and 5). MetaLibri Digital Library. Retrieved from https://www.ibiblio.org/ml/libri/s/SmithA_WealthNations_p.pdf

Alex, S. (2023, November 4). *Bitcoin vs ethereum consensus: PoW Vs PoS mechanisms explained.* Retrieved from https://www.doubloin.com/learn/bitcoin-vs-ethereum-consensus

Alistair, S., & Eleftherios, K.-K. (2020, June). *GRANDPA: A Byzantine finality gadget.* Retrieved from grandpa.pdf: https://github.com/w3f/consensus/blob/master/pdf/grandpa.pdf

Andreas, A. A., & Gurvinder, D. (2021). UNISWAP: Impermanent loss and risk profile of a liquidity provider. *Trading and Market Microstructure, 16.* Retrieved from https://arxiv.org/abs/2106.14404

Bernd, S. C., Doris, S., & Rowena, R. (2022). *Ethics of artificial IntelligenceCase studies and options for addressing ethical challenges.* Springer. Retrieved from https://link.springer.com/book/10.1007/978-3-031-17040-9

Buterin, V. (2014). *Ethereum: A next-generation smart contract and decentralized application platform.* ethereum.org. Retrieved from https://ethereum.org/669c9e2e2027310b6b3cdce6e1c52962/Ethereum_Whitepaper_-_Buterin_2014.pdf

Cambridge Center for Alternative Finance. (2022, January). *Evolution of network hashrate.* Retrieved from https://ccaf.io/cbeci/api/v1.2.0/download/mining_countries

CoinMarketCap. (2023, December). *Bitcoin (BTC) dominance.* Retrieved from https://coinmarketcap.com/charts/#What%20is%20Bitcoin%20(BTC)%20dominance?

Dániel, K., Márton, P., István, C., & Gábor, V. (2014). Do the rich get richer? An empirical analysis of the bitcoin transaction network. *PLoS One.* https://doi.org/10.1371/journal.pone.0086197

ethereum.org. (2024, January). *The merge.* Retrieved from https://ethereum.org/en/roadmap/merge/

Freeman, E. R., Harrison, J., Wick, A., Parmar, B., & Colle, S. (2001). *A stakeholder Approch to strategic management.* Blockwell.

Galbraith, J. K. (1952). *American capitalism: The concept of countervailing power.* (K. Nikawa, Trans.) Hakusuisya.

Hasan, K. (Spring 2021). *Market efficiency for Bitcoin.* Master Thesis NEKN02, p. 29. Retrieved from https://lup.lub.lu.se/luur/download?func=downloadFile&recordOId=9051578&fileOId=9051581

Jacques, F. (1999). *Multi-agent system: An introduction to distributed artificial intelligence.* JASSS. Retrieved from https://jasss.soc.surrey.ac.uk/4/2/reviews/rouchier.html

Joseph, S. A. (1939). *Business cycles.* McGraw-Hill Book Company. Retrieved from https://discoversocialsciences.com/wp-content/uploads/2018/03/schumpeter_businesscycles_fels.pdf

Karl, P. (1978). Three worlds. *The Tanner lecture on human values*, 1–27. Retrieved from https://tannerlectures.utah.edu/_resources/documents/a-to-z/p/popper80.pdf

Kiyoshi, M., & Yohko, O. (2021). The privacy paradox: Invading privacy while protecting privacy. In *ETHICOMP 2021* (pp. 199–201). Universidad de La Rioja, Universidad Complutense Madrid, CCSR De Montfort University, CBIE Meiji University.

Nakamoto, S. (2008). *Bitcoin: A peer-to-peer electronic cash system.* Bitcoin.org. Retrieved from https://bitcoin.org/ja/bitcoin-paper

Nielsen. (2022, February). *The introduction to web 3.0 (NFT, DeFi, DAO, DApp, Cryptocurrency, GameFi, etc).* Retrieved from https://paragraph.com/@nelsen/the-introduction-to-web-3-0-nft-defi-dao-dapp-cryptocurrency-gamefi-etc

Omkar, G. (2023, April 12). *CoinDesk.* Bitcoin, Not Ether, Builds Crypto Market Dominance Ahead of Ethereum's Shanghai Upgrade. Retrieved from https://www.coindesk.com/markets/2023/04/11/bitcoin-not-ether-is-becoming-more-dominant-in-crypto-market-ahead-of-ethereums-shanghai-upgrade/

Perpetual Protocol. (2022, September). *medium.* Retrieved from What is an Automated Market Maker (AMM)?: https://medium.com/perpetual-protocol/what-is-an-automated-market-maker-amm-a71ea1d80ea9

Prateek, G. (2019). Bitcoinomics 101: Principles of Bitcoin's supply, demand & price. https://doi.org/10.2139/ssrn.3473279

Public. (n.d.). *Unchained capital.* Retrieved from Bitcoin age distribution: https://chart-studio.plotly.com/~unchained/37/bitcoin-utxo-age-distribution/#plot

Sal, M., (2019, January). *Comparison of PoW, PoS and DPoS governance models.* Retrieved from https://medium.com/@salmanmiah/comparison-of-pow-pos-and-dpos-governance-models-dcea481140f8

Shimizu, K. (2021). *Blockchain and biometrics authorization: What we actually count truly counts?* Retrieved from https://dialnet.unirioja.es/servlet/articulo?codigo=8037017

Shirer, M. (2023, December). *Worldwide semiconductor market outlook upgraded to GROWTH from TROUGH: Semiconductor market to grow 20.2% in 2024 to $633 billion, according to IDC.* Retrieved from https://www.idc.com/getdoc.jsp?containerId=prUS51383823

Shuai, W., Wenwen, D., Juanjuan, L., Yong, Y., Liwei, O., & Fei-Yue, W. (2019). Decentralized autonomous organizations: Concept, model, and applications. *IEEE Transactions on Computational Social Systems.* Retrieved from https://ieeexplore.ieee.org/abstract/document/8836488

Simon, R. (2023). *Technology ethics: The ethical digital technology trilogy.* Routledge.

Stanford. (2020, October). *Stanford encyclopedia of philosophy.* Retrieved December 2022, from Hegel's Dialectics: https://plato.stanford.edu/entries/hegel-dialectics/

Stephen, S. (2015). *Bitcoin: The Napster of currency* (pp. 581–641). J.D. University of Houston Law Center. Retrieved from https://scispace.com/pdf/bitcoin-the-napster-of-currency-40sy6u-jxe2.pdf

Uniswap. (2023, June). *Welcome to uniswap docs.* Retrieved from Order Book VS AMM: https://docs.uniswap.org/concepts/uniswap-protocol

WSTS. (2023, December). *Historical billings report.* Retrieved from https://www.wsts.org/esraCMS/extension/media/f/WST/6280/WSTS-Historical-Billings-Report-Oct2023.xlsx

Yamazaki, T., Murata, K., Orito, Y., & Shimizu, K. (2020). Post-truth society: The AI-driven society where no one is responsible. In *ETHICOMP 2020.* Universidad de La Rioja. Retrieved from https://www.researchgate.net/publication/343099318_Post-Truth_Society_The_AI-driven_Society_Where_No_One_Is_Responsible

Chapter 10
Beating Bitcoin

Abstract Using Bitcoin, often referred to as "digital gold (Proof of Work, PoW)," as a benchmark, this study investigates the development and promotion of ecosystems driven by "digital oil (Proof of Stake, PoS)." The findings can be summarized in three key points: (1) Revolution in Decentralized Governance. While current systems are dominated by centralized control, often referred to as the "Magnificent Seven," the Bitcoin network achieves decentralized data management and governance through consensus mechanisms like PoW and PoS. This shift enables a democratic revolution where ownership lies with the users. (2) Universal Laws of Success. The Bitcoin network exemplifies the mathematical "power law," representing universal principles of success. It quantifies economic growth dynamics exponentially, offering a hybrid perspective that combines traditional capital market investment strategies with approaches tailored to emerging (start-up) markets. This framework facilitates a paradigm shift in the economy by enhancing productivity. (3) Optimizing Resource Allocation. The current global economy relies on single value benchmarks, such as the US dollar, which leads to inefficiencies and resource misallocations. Adopting new benchmarks like Bitcoin, combined with alternative indicators (including altcoins), could enable better resource allocation and measurement. This process, driven by "tokenized time" as a new productivity metric, holds the potential to improve resource management and create a more efficient economic framework.

Keywords Bitcoin · Blockchain · Proof of work · Learning curve · Tokenized time · ESG · Industrie 4.0 · Power law

K. Shimizu, *ESG Investment, German Industrie 4.0, and Blockchain*, https://doi.org/10.1007/978-981-95-6927-4_10

10.1 Introduction

Bitcoin has surpassed the $100,000 mark. How can this price formation be theoretically justified? Much like the self-organizing behavior of an ant colony, the Bitcoin network operates autonomously. Since the emergence of Bitcoin (BTC), the cryptocurrency market has developed rapidly, giving rise to diverse projects and use cases. While this growth has presented investors with new opportunities, it has also introduced complex risks and dynamics that traditional investment methodologies cannot fully capture.

This study aims to investigate the transition from PoW to PoS, symbolizing a shift from "digital gold" to "digital oil," and explore how PoW governance evolves into a PoS governance-based ecosystem. This research focuses on the structural changes and the promotion and dissemination of governance functions within this context. From this perspective, Bitcoin is positioned as a benchmark, and the following analyses are conducted (Galbraith, 2010).

First, the study highlights the relationships and differences between Bitcoin and other cryptocurrencies, establishing Bitcoin as a benchmark.

Second, it provides a theoretical understanding of Bitcoin's network structure by incorporating Google's PageRank and Smith's concept of empathy, thereby explaining its value from both a technological and economic perspective.

Finally, the study applies the statistical power law model to the concepts of PoW and decentralized networks. This approach demonstrates that real-world phenomena often follow a "long tail" distribution, where a few top entities dominate while the distribution tapers off into a smooth curve toward the right (Dániel et al., 2014).

10.2 Theoretical Background

Shannon mentions that the interaction between noise and signal in Shannon's communication model is analogous to Bitcoin's network structure, including mining, trading volume, and the balance of supply and demand (Shannon, 1949). This suggests that an underlying order emerges despite the apparent randomness in Bitcoin's price movements, highlighting a self-organizing system within the market.

10.2.1 ESG Investment

The relationship between ESG investment and Bitcoin analyzed through the lens of network theory highlights how Bitcoin as a monetary network compares with traditional companies in terms of growth trajectory and scale, such as Amazon and the Magnificent Seven companies for destroying the middleman's jobs. This comparison provides insights into Bitcoin's potential role within sustainable investment frameworks like ESG.

Bitcoin reached a valuation comparable to $1 trillion within 12 years, significantly faster than traditional companies:

- Microsoft: 44 years
- Amazon: 24 years
- Apple and Google: 22 years

This rapid growth reflects Bitcoin's network effect, where the value of the network increases exponentially as more participants (users, miners, and institutions) adopt and contribute to it. In network theory, this is known as Metcalfe's Law: the value of a network is proportional to the square of the number of its users (Michael, 2021).

Bitcoin's growth demonstrates that network effects can drive rapid value creation, making it an attractive asset within a diversified ESG portfolio if managed sustainably. Comparing Bitcoin to traditional networks like Microsoft and Amazon highlights its potential to redefine value networks while addressing challenges such as energy efficiency and regulatory compliance.

Bitcoin's rapid rise to a trillion-dollar monetary network reflects its unique position as a decentralized, scalable, and transparent system. When evaluated through network theory and aligned with ESG principles, Bitcoin represents both an opportunity and a challenge for sustainable investment strategies. It emphasizes the potential for technological innovation to drive both economic and social value, reshaping traditional financial paradigms (Cambridge Centre for Alternative Finance, 2021).

10.2.2 Industrie 4.0

Industrie 4.0 represents the fourth industrial revolution, characterized by integrating Cyber-Physical Systems (CPS) and the Internet of Things (IoT) into manufacturing and production processes (acatech POSITION PAPER 2011). German and Japanese manufacturing industries, particularly in the automotive sector, focus heavily on modular production systems (e.g., Volkswagen (VW)'s MQB[1] and Toyota's TNGA[2]) to enhance efficiency, reduce costs, and improve supply chain management (SCM) . German and Japanese manufacturing industries, particularly in the automotive sector, focus heavily on modular production systems (e.g., VW's MQB and Toyota's TNGA) to enhance efficiency, reduce costs, and improve supply chain management (SCM) (Shimizu, 2016a; Shintaku, 2016).

Smart Supply Chains: By integrating blockchain with IoT, manufacturers can create intelligent supply chains where every component's journey is recorded and verified in real time (Shimizu, 2021). Predictive Maintenance: Blockchain can log

[1] MQB; Modulare Quer Baukasten (in German), Modular Transverse Matrix in English.

[2] TNGA; Toyota New Global Architecture.

sensor data from CPS systems, enabling predictive maintenance models that minimize downtime and optimize machine usage. Decentralized Energy Grids: As smart grids expand (part of Industrie 4.0), blockchain can facilitate decentralized energy transactions, enabling efficient and transparent management of energy resources (Toyota Blockchain Lab, 2024).

Industrie 4.0 and IoT revolutionize manufacturing by creating interconnected, data-driven ecosystems. Blockchain networks complement this transformation by providing secure, transparent, and decentralized solutions for data management and process automation. The synergy between these technologies is essential for achieving greater efficiency, innovation, and sustainability in modern industrial systems.

10.2.3 PageRank

PageRank, an algorithm developed by Google, ranks web pages based on the quantity and quality of links pointing to them. Links from high-quality pages confer more "importance," reflecting collective preferences, much like Adam Smith's "impartial spectator" concept in economics (Shimizu, 2014) (Adam, 2007). Similarly, Bitcoin operates as a decentralized network where nodes (users, miners, transactions) derive value from trust and connectivity (Sergey & Lawrence, 1998).

- Network Effects: Both PageRank and Bitcoin leverage network effects, where value increases with more quality connections. In Bitcoin's case, its network becomes stronger and more secure as participants grow (Shimizu, 2016b).
- Decentralization: While PageRank relies on decentralized data inputs, it remains centrally governed by Google, Bitcoin network achieves true decentralization in data and governance through consensus mechanisms like Proof of Work (or Proof of Stake) (Gilder, 2018).
- Value Creation: PageRank demonstrates how networks create value through influence and connectivity. Bitcoin mirrors this, deriving its value from widespread trust and adoption within its decentralized framework. Well-connected nodes, such as large miners and active wallets, play crucial roles, akin to high-PageRank pages.
- The Power of Democratization: Both systems demonstrate the power of network theory in different domains: PageRank in the field of information retrieval and Bitcoin in monetary systems. In particular, Bitcoin's rapid adoption has increasingly positioned it as a powerful force for transformation, working to democratize monetized and entrenched networks in marketing and to rebuild stagnating modern digital ecosystems.

In essence, PageRank's ranking logic and Bitcoin's network dynamics reflect a shared reliance on trust, connections, and decentralization to generate influence and value. However, unlike Google's PageRank, which a private corporation controls, Bitcoin's decentralized structure prioritizes individual ownership and transparency, which is done by Intimax using Plasma technology. This shift highlights a vision for a Web3 era, where individuals, as information contributors, leverage decentralized infrastructure to maintain ownership and privacy, moving away from centralized control and toward a more user-centric and equitable digital ecosystem (Shimizu, 2014).

10.2.4 Network and Success

"The Formula" to the Bitcoin Network applies the several universal laws of success outlined in The Formula: The Universal Laws of Success by Albert-László Barabási to the Bitcoin network, highlighting its growth dynamics and the factors contributing to its success below (Albert, 2018).

10.2.4.1 Performance vs. Network Power

Bitcoin's success is not determined solely by its technical performance (e.g., transaction speed or security) but heavily depends on its network effects. The increase in users, miners, and transaction volumes enhances Bitcoin's value, exemplifying Metcalfe's Law. Additionally, the participation of institutional investors as network participants has become a significant factor in its growth.

10.2.4.2 Performance Limits, Boundless Success

While Bitcoin's technical performance has limitations (e.g., transaction capacity), its success as a monetary network is unbounded. Its price growth, from a few dollars to tens of thousands, demonstrates how demand and recognition can transcend technical constraints, further solidified by its adoption in financial markets (e.g., ETFs and institutional investments). For example, examining the relationship between long-term stock indices and time (such as the S&P 500 without reinvested dividends) reveals that major stock market indices are widely known to grow at an average exponential rate of approximately 7% per year.

10.2.4.3 Persistence and Timing

Bitcoin's success highlights the importance of persistence. Despite initial skepticism, Bitcoin has endured over a decade, evolving through technological advancements (e.g., Lightning Network) and addressing challenges (e.g., scalability and energy use). Its timing has also been critical, with growing interest in decentralized finance and digital assets.

By applying Barabási's framework, Bitcoin's success can be understood as a result of network effects, adaptability, persistence, and the convergence of societal trust and technological innovation. These insights not only clarify Bitcoin's unique position as a decentralized monetary network but also offer a model for evaluating other digital assets in the context of network-driven growth and success.

10.3 Compute-Speed and Power Law

The significant price movements in Bitcoin, akin to "black swan events" in power-law distributions, underline the idea that a few extreme outcomes dominate the landscape. In Barabási's terms, these are the "unbounded" aspects of success, where Bitcoin's value is not capped by physical constraints but by market dynamics.

10.3.1 Implications for Market Behavior

Understanding Bitcoin's price in terms of power laws allows for better anticipation of "rare but significant" price events. This is similar to how Barabási analyzes scientific breakthroughs or social phenomena, emphasizing the importance of looking for patterns in seemingly chaotic data.

In summary, Bitcoin's price dynamics reflect power-law behavior, shaped by network effects and collective perception. This aligns with Barabási's principles, where success (price movements) is driven not just by inherent performance but by the broader response of the community and market networks. In other words, Bitcoin as "digital gold" can serve as a benchmark for other governance systems and functions as a source of authority and resource provision for the construction of an ecosystem.

10.3.1.1 Super Generalized Central Limit Theorem (SGCLT)

Here, we will refer to the Super Generalized Central Limit Theorem (SGCLT), which is an extension of the "Generalized Central Limit Theorem" in statistics, in order to explore the possibility of diversification in the crypto market (Shintani & Umeno, 2018). This theory explains how data with different characteristics can be aggregated into a stable form as a whole, and is particularly suitable for data that is highly volatile and has a variable distribution, such as tokens in the crypto market.

The application of SGCLT enables the following new approaches:

Leveraging the characteristics of non-identical random variables enables the design of diversified portfolios. By appropriately combining tokens with varying volatility and distribution properties, it is possible to enhance the overall stability of the portfolio while maximizing returns. For example, between 2019 and 2020, many altcoins significantly outperformed Bitcoin. This demonstrates the potential to capture market growth in a risk-adjusted manner by utilizing the characteristics of tokens based on power-law distributions. In this context, the relationship established with Bitcoin provides valuable insights.

By deeply understanding the distribution properties of tokens (which are also related to the nature of their underlying projects), it becomes possible to model market dynamics that traditional investment theories cannot fully explain, enabling more proactive investment decisions. Historical performance shows that since 2021, only 10–20% of the top 150 tokens have outperformed Bitcoin, with those tokens achieving an average outperformance of approximately +100%. Considering the power-law distribution and diversity, a small number of high-performing tokens disproportionately impact the overall portfolio returns.

Investors can benefit from a venture capital-style diversification approach, effectively managing risks while leveraging the advantages of highly liquid secondary markets. This approach offers a hybrid perspective that combines traditional capital market investment methods with strategies for investing in emerging (start-up) markets.

10.3.2 Power-Law Distributions in Bitcoin Price

If Bitcoin can be seen as a system that "combines traditional capital market investment methods with strategies for investing in emerging (start-up) markets," then its price reflects the characteristics of small market capitalization within the ecosystem, being heavily influenced by infrequent but significant price fluctuations. This aligns with the behavior of emerging industries, which exhibit large variations following a power-law distribution. Here, Bitcoin can be positioned as a newcomer in the capital market, offering a perspective that applies the principles of power-law distribution (Burger & Vijn, 2024).

According to the IMF, while the total market capitalization of the global capital market is $36 trillion, the cryptocurrency market has reached $3.32 trillion, approximately 10% of the global scale. In this context, the fact that the price volatility in the cryptocurrency market is exceptionally high raises an important question: What does this imply? Could it serve as an indicator of potential in emerging enterprises, and how can this indicator be justified?

Power Law;

$$P_t \sim t^{-\alpha} \tag{1}$$

10.3.3 BTC-USD Close Price (Log Scale) by R

BTC-USD was obtained from Yahoo Finance US using crypto_tickers = ["BTC-USD"] with Python and a specified acquisition period (start_date = "2010-01-01"), and approximate values were calculated using both Python and R ($n = 3753$).

This graph represents Bitcoin's price and time on a logarithmic scale. Vertical Axis: The logarithmic transformation of Bitcoin's closing price (log(Price)). Horizontal Axis: The logarithmic transformation of the days elapsed since the first halving (November 28, 2012) (log(Time)).

The logarithmic scale on the vertical axis illustrates the changes in Bitcoin's price in an exponential scale. The logarithmic scale on the horizontal axis accounts for the time elapsed after the halving event, displayed in a logarithmic format. The points on the graph represent Bitcoin's price at specific dates, showing how the price evolves over time. This perspective helps to reveal the long-term trends in Bitcoin price fluctuations (see in Fig. 10.1).

Model overview for linear regression results (Log(Price) ~ Log(Time)): Perform linear regression with Log(Price) as the dependent variable and Log(Time) as the independent variable and analyze the residuals.

- R-squared (coefficient of determination): 0.759 → Log(Time) explains Log(Price) by about 75.9%.
- F-statistic: 1.18e+04 → The entire model is very statistically significant (P-value = 0.00).
- Coefficient: Const: −2.7381 → The intercept of the regression equation.

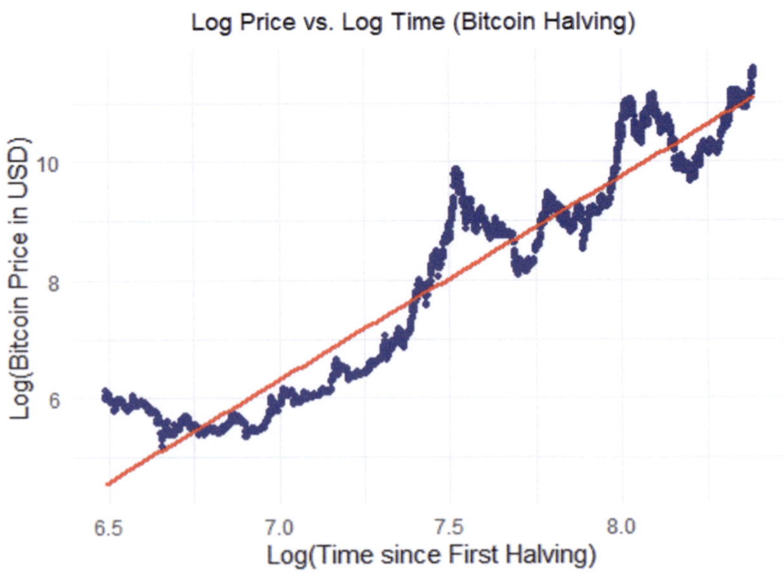

Fig. 10.1 BTC-USD close price (log scale), (Created by the author)

- Log_Time: 1.5973 → Log(Price) has a positive correlation with Log(Time).
- Statistical significance: Both coefficients have a *P*-value of 0.000 ($P < 0.05$), which is statistically significant. (Blei et al., 2003).
- Durbin-Watson statistic: 0.003 → suggests that there may be autocorrelation between the residuals.
- Regression equation: Log(Price) = −2.7381 + 1.5973 × Log(Time).

10.3.3.1 ADF (Augmented Dickey-Fuller) Test Results

- Test Statistics: −3.075
- *P*-value: 0.028
- Conclusion: The residuals are determined to be stationary (*P*-value <0.05).

Conclusion: A cointegration relationship exists between (Price) and Log(Time). The results suggest that Log(Price) and Log(Time) maintain a stable relationship in the long term, even if there are short-term fluctuations.

10.4 Money Is Tokenized Time

According to Marx's labor theory of value (lerning curve), the actual price of a good or service corresponds to the amount of labor time required to earn the money needed to purchase it. In other words, when we spend money, we are utilizing a value that has been indexed to our own time. For example, in a study by Nordhaus examining the amount of labor needed to afford 1 h of light for reading, the following results were observed (Gilder, 2023).

- Ancient Babylon (circa 1800–1500 BCE): Sesame oil lamps required 50 h of labor.
- Industrial Revolution (circa 1800 CE): Tallow candles required 6 h of labor.
- Incandescent bulbs (post-World War II, 1950): 8 s of labor.
- LED lamps: Minimal labor time to afford light (Nordhause, 1998).

These findings illustrate that innovations related to "light" follow a power-law distribution, with labor time decreasing exponentially as technology advances.

When applying this relationship to Bitcoin as a driver of innovation, Smith's concept of key drivers—"division of labor," "exchange," and "desire"—provides a compelling framework. In this context, the time indexed as value is maximized through actions that pursue economic efficiency, thereby translating into the modern value metric: money (Shimizu, 2022).

For example, in the case of light, if the labor required to provide it decreases exponentially, efficiency improves, and prices can rise as the value of the service increases. Similarly, goods and services with high diffusion and appeal can lead to broader economic efficiency through innovation. If the inefficiencies of using a single value medium such as the current global reserve currency (e.g., the US dollar) result in resource misallocation, adopting a new benchmark like Bitcoin, combined with alternative indicators (including altcoins), could enable better resource allocation and measurement.

Altcoins such as Ethereum and others such as Polkadot could complement Bitcoin by addressing its limitations, forming a new ecosystem that leverages innovation and diversity. This hybrid approach holds the potential to improve economic outcomes, such as optimizing energy use and reducing transaction costs, thereby aligning with broader efficiency and sustainability goals.

10.4.1 From "Time Is Money" to "Money Is Time"

The phrase "Time is Money" has long symbolized modern capitalism. However, with technological progress, an inverted idea—"Money is Time"—is now emerging as a new paradigm.

According to a historical study, the number of working hours required to obtain 1 h of light for reading has dramatically decreased over time. In ancient Babylonia, using a sesame oil lamp required about 50 h of labor. During the Industrial Revolution, candles required about 6 h; after World War II, the light bulb required only 8 s; and with today's LED lamps, the cost is virtually zero. This shift represents not merely an improvement in energy efficiency but the historical liberation of human time itself.

In his lecture "Can We Simulate Nature with a Computer?", physicist Richard Feynman argued that while classical physics is deterministic and reproducible, quantum mechanics is probabilistic, making it impossible for traditional computers to simulate exactly. He claimed that "to simulate quantum systems, we must use quantum systems," thus proposing the concept of the quantum computer. This idea marked a turning point in viewing nature and time through a new technological lens.

Economist William Nordhaus pointed out that conventional indicators such as wages and GDP fail to account for the "time-saving effects" brought about by technological innovation. His insight redefines wealth not as the accumulation of money, but as the saving of time and the expansion of freedom. Likewise, economic thinker George Gilder argues that time itself is becoming tokenized—a tradable form of capital. In financial terms, this suggests that the relationship between time and return is transforming, as returns from traditional investments are gradually being exchanged for the higher growth potential of new technologies such as blockchain.

Blockchain technology embodies this concept by tokenizing diverse forms of value and integrating ethical dimensions—such as ESG investment and Effective Altruism (EA)—into economic networks. For instance, it enables new mechanisms for converting inefficient market transactions into transparent, exchangeable systems. Technology, therefore, is not merely a tool for efficiency, but a means for humanity to reclaim time as a resource. The essence of wealth is shifting toward a new standard—how much time one can freely use, rather than how much money one can earn.

10.5 Conclusion

This paper provides a comprehensive analysis of Bitcoin's unique position within the cryptocurrency ecosystem, highlighting its dual role as a decentralized monetary network and a transformative asset. Through a comparative examination of Proof of Work (PoW) and Proof of Stake (PoS), Bitcoin is established as a benchmark cryptocurrency, showcasing its distinct technological and economic foundations. The investigation into Bitcoin's techno-economic value within ESG investment and Industrie 4.0 frameworks underscores its potential to drive sustainable innovation while addressing energy efficiency and regulatory compliance challenges.

Bitcoin's price dynamics, analyzed using power-law frameworks and cointegration theory, reveal its alignment with network-driven growth and market behavior. These findings emphasize the importance of network effects, as illustrated by Metcalfe's Law, in fostering rapid adoption and exponential value creation. Furthermore, the integration of traditional capital market investment methods with strategies tailored for emerging markets demonstrates a hybrid perspective that aligns Bitcoin with modern investment paradigms.

Unlike centrally governed systems such as Google's PageRank, Bitcoin achieves true decentralization in data and governance through consensus mechanisms like Proof of Work and Proof of Stake. This decentralized structure enhances its scalability, transparency, and resilience, positioning Bitcoin as a foundational element of the Web3 era. In this decentralized framework, individuals maintain ownership and privacy, reshaping traditional financial systems into more user-centric and equitable digital ecosystems.

Ultimately, Bitcoin's transformative potential lies in its ability to redefine economic and social value through innovation, while providing a benchmark for evaluating other digital assets within a rapidly evolving global financial framework. This study underscores the necessity of viewing Bitcoin not just as a speculative asset but as a cornerstone for future sustainable and efficient financial systems.

10.6 Review Questions

1. Explain the difference between "digital gold" (Proof of Work) and "digital oil" (Proof of Stake).
2. Describe how the centralized system represented by the "Magnificent Seven" differs from Bitcoin's decentralized governance structure.
3. Explain the economic and social impacts of the "network effect."
4. How do Bitcoin's energy efficiency and regulatory compliance issues pose challenges for sustainable investment strategies?
5. How can blockchain technology contribute to decentralized energy management, such as smart grids?

6. What is the "power law," and how is it related to Bitcoin's price movements?
7. How does the concept "Money is Time" differ from the traditional capitalist idea "Time is Money"?

References

acatech POSITION PAPER. (2011). *Cyber-physical systems: Driving force for innovation in mobility, health, energy and production*. Retrieved on 12. 9. 2017 von http://www.acatech.de/fileadmin/user_upload/Baumstruktur_nach_Website/Acatech/root/de/Publikationen/Stellungnahmen/acatech_POS

Adam, S. (2007). *An inquiry into the nature and causes of the wealth of nations* (Vol. 1, 2, 3, 4 and 5). MetaLibri Digital Library. Retrieved from https://www.ibiblio.org/ml/libri/s/SmithA_WealthNations_p.pdf

Albert, B.-L. (2018). *The formula: The universal laws of success.* Little, Brown and Company. Retrieved from https://www.academia.edu/75141375/Albert_L%C3%A1szlo_Barab%C3%A1si_The_Formula_The_Universal_Laws_of_Success_The_Science_Behind_Why_People_Succeed_or_Fail

Blei, D. M., Ng, A. Y., & Jordan, M. I. (2003). Latent Dirichlet allocation. *Journal of Machine Learning Research*. Retrieved from https://jmlr.org/papers/volume3/blei03a/blei03a.pdf

Burger, H. C., & Vijn, P. (2024, January 31). *Medium.* Retrieved from Bitcoin's time-based power-law and cointegration revisited: https://medium.com/quantodian-publications/bitcoins-time-based-power-law-and-cointegration-revisited-15f212b95ea5

Cambridge Centre for Alternative Finance. (2021). *Bitcoin network power demand.* Retrieved from https://cbeci.org/

Dániel, K., Márton, P., István, C., & Gábor, V. (2014). Do the rich get richer? An empirical analysis of the bitcoin transaction network. *PLoS One*. https://doi.org/10.1371/journal.pone.0086197

Galbraith, J. K. (2010). *The affluent society and other writings 1952–1967*. Literary Classics of the United States.

Gilder, G. (2018). *Life after Google: The fall of big data and the rise of the Blockchain economy*. Simon and Schuster.

Gilder, G. (2023). *Life after capitalism: The meaning of wealth, the future of the economy, and the time theory of money*. Regnery Gateway.

Michael, S. (2021). *Can bitcoin be replaced or fail?* Retrieved from https://www.youtube.com/watch?v=LdwSxwl9BQk

Nordhaus, W. D. (1998). *Do real-output and real-wage measures capture reality? The history of lighting suggests not*. Retrieved from http://www.nber.org/chapters/c6064

Sergey, B., & Lawrence, P. (1998, April). *The anatomy of a large-scale hypertextual web search engine.* Retrieved December 2024, from http://ilpubs.stanford.edu:8090/361/

Shannon, C. E. (1949). A mathematical theory of communication. *Bell System Technical Journal*. Retrieved from https://people.math.harvard.edu/~ctm/home/text/others/shannon/entropy/entropy.pdf

Shimizu, K. (2014). Governing "invisible hand" by cybernetics approach: Understanding Adam Smith's legacy; How "Google" create a new way of thinking by the influence of privacy. *IFSAM2014 World Congress in Tokyo*.

Shimizu, K. (2016a). *Smart factory or smart car? The concept of IoT usability.* Retrieved 2024, from https://www.researchgate.net/publication/341883267_Smart_Factory_or_Smart_Car_The_concept_of_IoT_usability

Shimizu, K. (2016b). Socio-cybernetic approach into the triumvirate: stakeholder governance between management, shareholders and employees. *An Enterprise Odyssey. International Conference Proceedings*, 273–280. Retrieved from https://search.proquest.com/docview/1815362010?pq-origsite=gscholar&fromopenview=true

Shimizu, K. (2021). *Blockchain and biometrics authorization: What we actually count truly counts?* Retrieved from https://dialnet.unirioja.es/servlet/articulo?codigo=8037017

Shimizu, K. (2022). Bullshit blockchain mining? *Proceedings of the ETHICOMP 2022.*

Shintaku, J. (2016). *Development of offshore production and its impact on domestic manufacturing*. Global Business Research Center: Akamon Management Review. https://doi.org/10.14955/amr.150501

Shintani, M., & Umeno, K. (2018). Super generalized central limit theorem—Limit distributions for sums of non-identical random variables with power laws. *Journal of the Physical Society of Japan, 87*(4). https://doi.org/10.7566/JPSJ.87.043003

Toyota Blockchain Lab. (2024, July 19). *How to introduce mobility into the public blockchain.* Retrieved December 2024, from https://www.toyota-blockchain-lab.org/library/how-to-introduce-mobility-into-the-public-blockchain

Chapter 11
Designing Time: ESG Investment, Blockchain, and the Convergence of Mobility and Energy in the Digital Era

Abstract This study aims to clarify how technological innovation, ESG investment, and digital governance are reconstructing the foundations of capitalism through the convergence of time, energy, and information. Building on George Gilder's concept that "money is tokenized time," it redefines the traditional notion of "time is money" into "money (tokens) is time," interpreting capital as stored and exchangeable time and presenting a new relationship between productivity, sustainability, and ethical value creation. Theoretically, it adopts the perspective of Actor–Network Theory (ANT) to analyze interrelationships and empirically examines the interactions among ESG investment, Industrie 4.0, and blockchain technology through statistical analysis and case studies.

A comparative regional analysis revealed that ESG investment, supported by strong policy frameworks and infrastructure, has accelerated the spread of electric vehicles (EVs) in Europe and the United States, while Japan, Canada, and Australia exhibit weaker linkages due to fragmented capital flows and delayed institutional adaptation. Furthermore, case studies such as TRENDE's blockchain-based P2P electricity trading and the BIS "Project Genesis 2.0" demonstrate how blockchain and IoT technologies enhance the transparency of carbon-credit systems and decentralized energy markets.

These findings indicate the emergence of an "information-value economy," in which environmental data, renewable energy, and financial instruments are tokenized. This model of decentralized network capitalism transforms sustainability from a moral ideal into a data-driven architecture of value co-creation, presenting a new paradigm that integrally reshapes technological innovation, financial systems, and social structures.

Keywords Tokenized time · ESG investment · Blockchain and transparency · Mobility–energy convergence · Information-value economy

K. Shimizu, *ESG Investment, German Industrie 4.0, and Blockchain*,
https://doi.org/10.1007/978-981-95-6927-4_11

11.1 Introduction

This study begins with George Gilder's perspective that "*money is tokenized time,*" and reconsiders the relationship between technological innovation and economic value. In contrast to Northous's idea of "*the labor time required for one hour of reading,*" improvements in productivity enhance quality of life by reducing working hours, thereby redefining capital as "*stored and exchangeable time.*" This view represents a shift from "*Time is Money*" *to* "*Money is Time,*" suggesting that the ethical returns of ESG investment can be connected and exchanged with other value networks, such as carbon credits (Albert 2018).

At the same time, electrification (EV) and digital transformation (DX) in the automotive industry are profoundly changing consumer behavior and awareness of sustainability. With the introduction of IoT and AI and under fluctuating energy prices, consumers are becoming more focused on cost-effectiveness across the entire life cycle and making more rational and sustainable decisions.

This project interprets these structural changes as the interaction between ESG investment, Industrie 4.0, and blockchain technology, exploring new design principles for sustainable economic systems. It does so through both theoretical frameworks and empirical case studies, such as home solar EV integration and green finance innovation.

11.2 Actor–Network Theory (ANT)

Adam Smith and Karl Marx argued that the Reformation, driven by Gutenberg's printing press, replaced manual copying with mass reproduction, increased literacy, and facilitated the widespread dissemination of ideas and concepts. Likewise, the Industrial Revolution, exemplified by the spinning jenny powered by the steam engine, illustrated how technology—through time—acquires autonomous momentum, reshaping social structures (Smith & Marx, 1995). The instructional structure for operating the complex warp and weft of weaving machines during the Industrial Revolution became the basis for punched-card systems, which later served as the foundational technology for computer memory (Deutsches Museum, 2022).

In the network society shaped by the Internet—a product of modern information and communication technologies—technology likewise gains autonomous momentum over time, continuously restructuring social organization (Hall, 1994). In this sense, "*time*" itself becomes flexible and transformable through technology (Gilder, 2023).

Network theory, broadly defined, spans fields such as communication, computation, biology, and sociology, and analyzes structures composed of nodes and edges using mathematical and generalized models. In this study, we adopt Actor–Network Theory (ANT) to analyze networks that include not only human but also non-human "actors," examining their mutual interrelations (Morgan, n.d.). Within such a networked society, the existence and credibility of each actor are secured through systems like Satoshi Nakamoto's blockchain design, which addressed the double-spending problem (the impossibility of duplication or imitation) by encrypting information and broadcasting it to distributed nodes (Nakamoto, 2008). This theoretical framework interprets contemporary technologies—such as blockchain and AI—as embedding within themselves certain values (e.g., tendencies toward decentralization or algorithmic decision criteria) that exert influence over human decision-making processes and institutional design.

11.3 ESG Investment as a Network of Nodes

According to estimates by Deutsche Bank, the total value of global financial assets in 2014 was approximately $242 trillion (Deutsche Bank Research, 2014). Statistics from the World Federation of Exchanges (WFE) in 2017 show that the total value of equities ($85.3 trillion), bonds ($76.2 trillion), and derivatives ($8 trillion) amounted to about $169.5 trillion in securities assets (World Federation of Exchanges, n.d.). Globally, the latest estimate by SIFMA (2024) indicates that the outstanding value of bonds has expanded to approximately $145 trillion (2025-SIFMA-Capital-Markets-Factbook, 2025).

According to the Global Sustainable Investment Review 2022 sustainable investment assets amount to approximately $30.3 trillion (The Global Sustainable Investment Alliance, 2025). By region, the composition is as follows: Europe—46%, United States—28%, Australia and New Zealand—8%, Japan—14%, and Canada—4%. The notable decline in the US share is attributed to the implementation of stricter measures aimed at preventing greenwashing.

Regarding sustainable bonds (such as green bonds), the Climate Bonds Initiative (CBI) reports that by the end of 2024, the cumulative issuance of GSS (Green, Social, and Sustainability bonds) and SLB (Sustainability-Linked Bonds)—collectively referred to as GSS+—has reached USD 6.9 trillion, of which USD 5.7 trillion (83%) corresponds to GSS+ instruments (Climate Bonds Initiative, 2025).

Of the $145 trillion in global outstanding bonds mentioned above, green bonds accounted for approximately 4.7%. The share of sustainable investments (valued at $30.3 trillion) represented about 21% of the total. This trend indicates that the integration of ethical and environmental perspectives into investment decision-making is becoming mainstream.

However, the proportion of green bonds within the overall bond market remains relatively small—around 5%—showing that it is still in a developing stage. Nevertheless, led particularly by Europe, regulatory frameworks such as the EU Taxonomy and mandatory ESG disclosures are advancing, thereby mainstreaming corporate and governmental financing through sustainability-linked instruments.

This phenomenon implies that a structural transformation is underway—one that incorporates emerging technologies such as blockchain to enhance transparency, traceability, and accountability in sustainable finance.

11.3.1 Correlation Analysis Between EV Sales and ESG Investment

This study conducted a statistical analysis of the relationship between electric vehicle (EV) sales and ESG investment using data from the International Energy Agency (IEA) and the Global Sustainable Investment Alliance (GSIA), see Fig. 11.1 below. ESG investment figures from GSIA are reported separately for Europe, the United States, Canada, Australia & New Zealand, and Japan. These figures were converted to US dollars and analyzed alongside IEA data using natural logarithmic transformation.[1]

The results revealed distinct patterns when dividing the regions into two groups:

- Europe/US (Regression Coefficient **β = 1.46**): In this group, a one-unit increase in the natural logarithm of ESG investment corresponds to a 1.46-unit increase in the log of EV sales.

This elasticity greater than one indicates an accelerated effect, where ESG investments significantly boost EV adoption. The strong response is likely due to

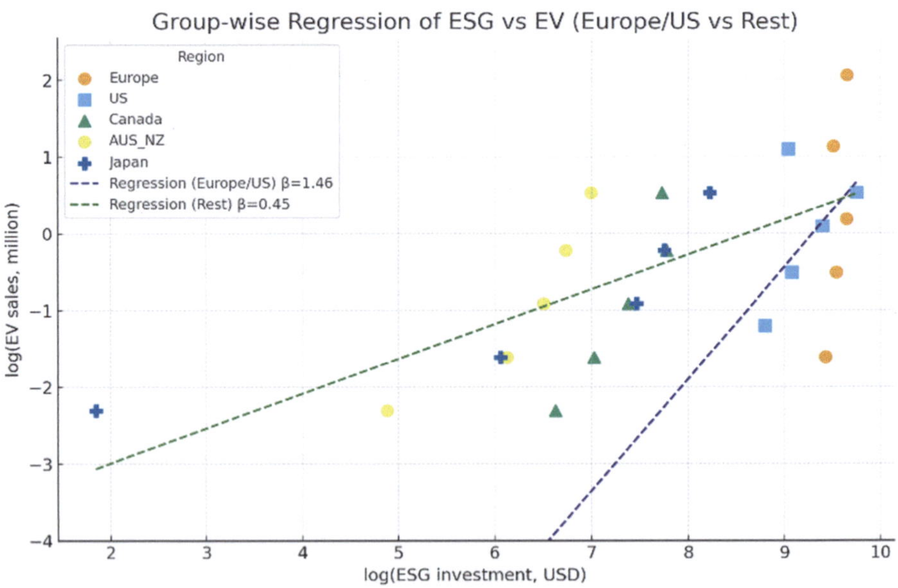

Fig. 11.1 Correlation between ESG investment (USD standardized) and EV sales (millions of car sales), (Created by the author, using R and Python)

[1] Asset values are expressed in billions. Exchange rate for average from 2014〜2022; Europe (EUR):USD = EUR × 1.25, United States (USD):USD = USD × 1.00, Canada (CAD):USD = CAD × 0.75, Australia & New Zealand (AUD):USD = AUD × 0.65, Japan (JPY):USD = JPY × 0.0075 (= 1 USD ≈ 133.33 JPY) and Log scale; log_ESG = ln(ESG_USD), log_EV = ln(EV_million), Model $\log(EV) = \alpha + \beta * \log(ESG) + \varepsilon$ (OLS)

concentrated investment in EV-related sectors, mature infrastructure, supportive policy frameworks, and robust supply chains.

- Rest of the World (Canada, AUS_NZ, Japan: Regression Coefficient **β = 0.45**) In contrast, these regions showed a more modest relationship: a one-unit increase in ESG investment was associated with only a 0.45-unit increase in EV sales (both in logarithmic terms).

This weaker elasticity suggests that ESG capital may be dispersed across various sectors—such as real estate, renewable energy, and forestry—and that EV-specific policy and infrastructure are relatively underdeveloped. In particular, Japan and Australia/New Zealand may be facing structural delays in integrating renewable energy with EV policy and infrastructure, which could limit the effectiveness of ESG investment in driving EV adoption.

These regional disparities in the impact of ESG investment on EV adoption can be attributed to a combination of interrelated structural, financial, and behavioral factors.

First, policy support plays a critical role. Europe and the United States have implemented strong regulatory frameworks that directly promote EV adoption, including generous subsidies, carbon pricing mechanisms, and emissions reduction mandates. In contrast, policy support in the Rest group—excluding China, the world's largest EV producer—is generally weaker and less stable. For instance, Japan has experienced inconsistent subsidy programs for EVs, undermining long-term investment confidence and consumer planning.

Second, there are notable differences in ESG capital allocation. In Europe and the United States, ESG funds are often strategically directed toward EV-related technologies such as electric drivetrains, battery innovation, and charging infrastructure. In the Rest regions, however, ESG investment is typically more fragmented, being allocated across a broader range of sectors including renewable energy, sustainable real estate, and forest conservation. This dispersion can dilute the direct impact of ESG funding on EV deployment.

Third, infrastructure development strongly influences outcomes. The EV ecosystems in Europe and the US benefit from more mature and standardized charging networks, enabling broader adoption. Conversely, in countries like Canada and Australia, vast land areas and uneven population distribution make it more difficult and cost-intensive to develop comprehensive EV infrastructure.

Finally, consumer behavior and cultural norms further contribute to the divergence. In Europe and the United States, environmental concerns and the economic appeal of EVs are increasingly embraced by consumers. By contrast, gasoline-powered vehicles remain the norm in many Rest countries, where cultural familiarity, fuel price structures, and market inertia continue to inhibit EV adoption.

Taken together, these factors help explain why the same level of ESG investment can yield significantly different outcomes across regions, with Europe and the United States experiencing accelerated EV growth while others remain at earlier stages of transition.

11.4 Mobility–Energy Convergence in the Digital Era

11.4.1 EVs as IoT-Enabled Digital Mobility

Modern electric vehicles (EVs) have evolved into software-centric products often described as "computers on wheels." Makers like Tesla keep vehicles permanently connected to the Internet; over-the-air (OTA) updates continue to enhance performance and functions after purchase. In this new generation of "smartphones on wheels," development is shifting from hardware-first to software-first, with product design that assumes cloud connectivity (with hyperscalers) and agile methods. As a result, frequent feature updates and security patches are permeating the auto industry, and EVs are rapidly maturing as true digital-mobility products (Kimberly et al., 2025).

11.4.2 EV–Home Solar Coupling: The Energy "Node" at the Edge

Because EVs run on electricity, coupling them with home solar PV yields strong energy synergies. Charging with self-generated solar lowers grid-purchased electricity costs, and driving the EV cuts gasoline expenses—creating a dual economic effect in post-FIT self-consumption models. In addition, EVs function as "mobile batteries"; with vehicle-to-home (V2H) systems, surplus daytime solar can be stored in the EV and supplied back to the home at night or during outages. In this way, combining EVs with rooftop PV circulates generation and consumption at home, advancing energy self-sufficiency. EVs thus serve not only as mobility devices but also as energy nodes that bridge the home and the grid, potentially boosting the combined ROI (Return on equity) of solar panels and EVs (IEA PVPS Task 17, 2021).

11.4.3 Prosumer Economics and New Expectations for ESG Investment

As EV–renewables integration progresses, consumers become prosumers—both producing and consuming energy. This shift is seen as essential to tackling the climate crisis: households generate clean power, use/store it via EVs, and sell or share surplus. Companies are responding with integrated offerings that combine EV charging, home storage, and solar control on unified platforms. Consumers gain economic benefits while contributing to decarbonization; firms create new revenue streams and customer value. From an ESG perspective, mobility–energy platform providers attract strong attention because their sustainable business models catalyze behavioral change (self-consumption, EV adoption) and can reinforce a virtuous cycle between environmental/social value and financial value (Shimizu, 2016). Incentive-driven prosumer strategies are emerging as a compelling theme for ESG investors (IEA, 2022).

11.4.4 Transparent Carbon-Credit Bonds and Tokenization: Toward an "Information-Value" Economy

Finance is also converging with mobility and energy. One example is smart-contract-based carbon-credit bonds.

11.4.4.1 Project Genesis 2.0

The Bank for International Settlements (BIS)–HKMA "Project Genesis 2.0" demonstrated tokenized green bonds with embedded mitigation-outcome interests (MOIs), linking bond outcomes to verifiable carbon credits and using IoT and blockchain for real-time impact tracking—reducing greenwashing risk (BIS, 2022). Hong Kong has since executed government digital/tokenized green bond issuances (multi-currency, digitally native) on the HSBC Orion platform via the HKMA's CMU system, advancing market infrastructure for such products. In parallel, tokenization and certificate platforms that use granular environmental data from distributed resources (e.g., PV generation, EV telemetry) are becoming practical.

11.4.4.2 TRENDE (Tepco's Subsidiary)

On September 6, 2024, TRENDE, in collaboration with the JA Group and ITOCHU Corporation, launched a peer-to-peer (P2P) electricity trading service utilizing blockchain and AI-based demand forecasting. The platform enables households equipped with solar panels and storage batteries to securely and transparently sell their surplus electricity directly to other consumers. By leveraging blockchain technology, transaction records and power traceability are managed in a tamper-proof manner, ensuring fairness and trust in local energy markets. The project aims to promote a "local production and local consumption" energy model by circulating post-FIT renewable power within communities. A pilot project is currently underway in Gunma Prefecture, with nationwide expansion to JA-affiliated facilities such as A-Coop planned in the near future (TRENDE, 2024).

11.5 Conclusion

This study has examined how technological innovation, financial systems, and ethical investment principles intersect to shape a new phase of capitalism—one where time, information, and energy become interlinked sources of value. Beginning from Gilder's notion that "money is tokenized time," we have explored how ESG investment, Industrie 4.0, and blockchain collectively redefine the relationship between productivity, sustainability, and trust.

Empirical analysis showed that ESG investment strongly correlates with EV adoption in Europe and the United States, where policy support, capital allocation, and infrastructure are aligned to accelerate decarbonization. In contrast, regions such as Japan, Canada, and Australia exhibit weaker linkages, reflecting fragmented ESG capital flows and underdeveloped policy frameworks. These disparities highlight that financial ethics alone cannot drive sustainability; technological and institutional infrastructures must co-evolve.

At the same time, the convergence of mobility and energy—illustrated by the integration of EVs, home solar systems, and V2H technologies—demonstrates how consumers evolve into prosumers, actively participating in energy production, storage, and distribution. Projects like TRENDE's blockchain-based P2P energy trading and BIS–HKMA's Project Genesis 2.0 exemplify this transition toward an information-value economy, where environmental data, carbon credits, and digital assets are tokenized to enhance transparency and accountability.

Ultimately, the synthesis of ESG investment, IoT-enabled industrial design, and blockchain governance suggests the emergence of a distributed economic model—one that transforms sustainability from a moral aspiration into an operational, data-driven architecture of value. In this "networked capitalism," time, energy, and information circulate through decentralized systems, creating feedback loops that align ethical intention with technological and financial innovation.

References

2025-SIFMA-Capital-Markets-Factbook. (2025, October). *2025-SIFMA-capital-markets-factbook.* Retrieved from https://www.sifma.org/wp-content/uploads/2024/07/2025-SIFMA-Capital-Markets-Factbook.pdf

Albert, B.-L. (2018). *The formula: The universal Laws of success.* Little, Brown and Company. Retrieved from https://www.academia.edu/75141375/Albert_L%C3%A1szlo_Barab%C3%A1si_The_Formula_The_Universal_Laws_of_Success_The_Science_Behind_Why_People_Succeed_or_Fail

BIS. (2022). *Genesis 2.0: smart contract-based carbon credits attached to green bonds.* Retrieved from https://www.bis.org/publ/othp58.pdf

Climate Bonds Initiative. (2025, October). *Global state of the market 2024.* Retrieved October 2025, from https://www.climatebonds.net/data-insights/publications/global-state-market-2024

Deutsche Bank Research. (2014). *Mapping the world's financial markets 2014*. Deutsche Bank. Retrieved from https://etf.dws.com/DEU/DEU/Download/Research-Global/47e36b78-d254-4b16-a82f-d5c5f1b1e09a/Mapping-the-World-s-Financial-Markets.pdf

Gilder, G. (2023). *Life after capitalism: The meaning of wealth, the future of the economy, and the time theory of money*. Regnery Gateway.

Hall, J. (1994). *Technological Determinism Summed Up*. Retrieved Oct 2025, from https://www.sccs.swarthmore.edu/users/97/jahall/thesis/tech/direct/determine.html#:~:text=%3E%20%60%20,once%20a,once%20a

IEA. (2022). *Unlocking the potential of distributed energy resources*. Retrieved from https://www.iea.org/reports/unlocking-the-potential-of-distributed-energy-resources?utm_source=chatgpt.com

IEA PVPS Task 17. (2021). *PV powered electric vehicle charging stations—Preliminary requirements and feasibility conditions*. Retrieved from https://iea-pvps.org/key-topics/pv-powered-electric-vehicle-charging-stations/?utm_source=chatgpt.com

Kimberly, B., Sebastian, K., Arturo, S., & Carolin, D. (2025). *Automotive product development: Accelerating to new horizons*. McKinsey. Retrieved from https://www.mckinsey.com/capabilities/operations/our-insights/automotive-product-development-accelerating-to-new-horizons?utm_source=chatgpt.com#/

Morgan, S. K. (n.d.). *Where are the missing masses? The sociology of a few mundane artifacts*. Retrieved Oct 2025, from https://www.morgan-klaus.com/readings/mundane-artifacts.html#:~:text=inherently%20anthropomorphic%2C%20in%20three%20ways%3A,to%20parts%20of%20humans%2C%20while.

Deutsches Museum (2022). Computers—The history of the calculating machine. Deutsches Museum. Abgerufen am Oct 2025 von https://www.deutsches-museum.de/en/museumsinsel/ausstellung/computers

Nakamoto, S. (2008). *Bitcoin: A peer-to-peer electronic cash system*. Bitcoin.org. Retrieved from https://bitcoin.org/ja/bitcoin-paper

Shimizu, K. (2016). *Smart factory or smart car? The concept of IoT usability*. Retrieved 2024, from https://www.researchgate.net/publication/341883267_Smart_Factory_or_Smart_Car_The_concept_of_IoT_usability

Smith, M. R., & Marx, L. (1995). *Does technology drive history? The Dilenuna of technological*. MIT Press. Retrieved from https://www.researchgate.net/publication/227622779_Does_Technology_Drive_History_The_Dilemma_of_Technological_Determinism_Merritt_Roe_Smith_Leo_Marx_Eds/link/64f16242c40f1d22df82c11d/download?_tp=eyJjb250ZXh0Ijp7ImZpcnN0UGFnZSI6InB1YmxpY2F0aW9uIiwicGFnZSI

The Global Sustainable Investment Alliance. (2025). *The global sustainable investment review 2022*. Retrieved from https://www.gsi-alliance.org/members-resources/gsir2022/

TRENDE. (2024). *TRENDE launches commercial P2P energy trading service in partnership with JA group and ITOCHU*. Retrieved from https://trende.jp/news/press/20240906_eng

World Federation of Exchanges. (n.d.). WFE Annual Statistics Guide (Volume 4). The United States. Retrieved from https://www.world-exchanges.org/our-work/articles/wfe-annual-statistics-guide-volume-4

Index

K. Shimizu, *ESG Investment, German Industrie 4.0, and Blockchain*, https://doi.org/10.1007/978-981-95-6927-4

MIX
Papier aus verantwortungsvollen Quellen
Paper from responsible sources
FSC® C105338

If you have any concerns about our products,
you can contact us on
ProductSafety@springernature.com

In case Publisher is established outside the EU,
the EU authorized representative is:
Springer Nature Customer Service Center GmbH
Europaplatz 3, 69115 Heidelberg, Germany

Printed by Libri Plureos GmbH
in Hamburg, Germany